Anneke Meyer

Local inter neurons in the antennal lobe of the honey bee

Anneke Meyer

Local inter neurons in the antennal lobe of the honey bee

A characterization based on single cell data

Südwestdeutscher Verlag für Hochschulschriften

Impressum/Imprint (nur für Deutschland/only for Germany)
Bibliografische Information der Deutschen Nationalbibliothek: Die Deutsche Nationalbibliothek verzeichnet diese Publikation in der Deutschen Nationalbibliografie; detaillierte bibliografische Daten sind im Internet über http://dnb.d-nb.de abrufbar.

Alle in diesem Buch genannten Marken und Produktnamen unterliegen warenzeichen-, marken- oder patentrechtlichem Schutz bzw. sind Warenzeichen oder eingetragene Warenzeichen der jeweiligen Inhaber. Die Wiedergabe von Marken, Produktnamen, Gebrauchsnamen, Handelsnamen, Warenbezeichnungen u.s.w. in diesem Werk berechtigt auch ohne besondere Kennzeichnung nicht zu der Annahme, dass solche Namen im Sinne der Warenzeichen- und Markenschutzgesetzgebung als frei zu betrachten wären und daher von jedermann benutzt werden dürften.

Verlag: Südwestdeutscher Verlag für Hochschulschriften GmbH & Co. KG
Heinrich-Böcking-Str. 6-8, 66121 Saarbrücken, Deutschland
Telefon +49 681 37 20 271-1, Telefax +49 681 37 20 271-0
Email: info@svh-verlag.de

Approved by: Konstanz, Universität Konstanz, Dissertation, 2011

Herstellung in Deutschland:
Schaltungsdienst Lange o.H.G., Berlin
Books on Demand GmbH, Norderstedt
Reha GmbH, Saarbrücken
Amazon Distribution GmbH, Leipzig
ISBN: 978-3-8381-1766-9

Imprint (only for USA, GB)
Bibliographic information published by the Deutsche Nationalbibliothek: The Deutsche Nationalbibliothek lists this publication in the Deutsche Nationalbibliografie; detailed bibliographic data are available in the Internet at http://dnb.d-nb.de.

Any brand names and product names mentioned in this book are subject to trademark, brand or patent protection and are trademarks or registered trademarks of their respective holders. The use of brand names, product names, common names, trade names, product descriptions etc. even without a particular marking in this works is in no way to be construed to mean that such names may be regarded as unrestricted in respect of trademark and brand protection legislation and could thus be used by anyone.

Publisher: Südwestdeutscher Verlag für Hochschulschriften GmbH & Co. KG
Heinrich-Böcking-Str. 6-8, 66121 Saarbrücken, Germany
Phone +49 681 37 20 271-1, Fax +49 681 37 20 271-0
Email: info@svh-verlag.de

Printed in the U.S.A.
Printed in the U.K. by (see last page)
ISBN: 978-3-8381-1766-9

Copyright © 2012 by the author and Südwestdeutscher Verlag für Hochschulschriften GmbH & Co. KG and licensors
All rights reserved. Saarbrücken 2012

Contents

1	**General Introduction.**		**1**
2	**Elemental and configural odour-coding by antennal lobe neurons of the honey bee.**		**5**
	2.1	Introduction	5
	2.2	Materials and Methods	7
		2.2.1 Animal preparation	7
		2.2.2 Odour stimulation	8
		2.2.3 Electrophysiology	8
		2.2.4 Morphology	9
		2.2.5 Data analysis	10
	2.3	Results	11
		2.3.1 Odour concentration can be used to enhance stimulus specific latency shifts in AL neurons.	11
		2.3.2 AL-neurons differ in their response patterns	12
		2.3.3 Elemental and configural coding both occur in AL neurons.	13
		2.3.4 AL-neurons are active sequentially.	16
		2.3.5 Hetero LNs are involved in configural as well as elemental processing.	17
	2.4	Discussion	18
		2.4.1 Broad and narrow tuned LNs - functional subgroups?	19
		2.4.2 Puzzles of suppression and excitation - a complex AL blueprint?	19
		2.4.3 Hetero LNs - multi-function neurons?	20
3	**Local Interneuron Morphology**		**23**
	3.1	Introduction	23
	3.2	Materials and Methods	25
		3.2.1 Animal preparation	25
		3.2.2 Stainings and morphological preparation	25
		3.2.3 Confocal imaging and data processing	27
	3.3	Results	28
		3.3.1 Inter-glomerular innervation patterns	28
		3.3.2 Intra-glomerular arborisations	29
		3.3.3 Branching shape of the main neurite.	30
		3.3.4 Morphological diversity of LNs	31
		3.3.5 Phenotype and neurite thicknes	33
	3.4	Discussion	33
		3.4.1 The LN attitude - PNcentric, or ORNcentric?	33

		3.4.2	Just homo or hetero? - LNs are morphologically diverse.	34
		3.4.3	Means of communication - possible assignment of neuro-transmitters and -peptides to the described LN phenotypes.	35

4 Clustering of evoked activity from antennal lobe neurons. 37

    4.1 Introduction 37

    4.2 Materials and Methods 38

        4.2.1 Data 38

        4.2.2 Data preprocessing 39

        4.2.3 Data descriptors 39

        4.2.4 Computation of descriptors 40

        4.2.5 Statistical analysis. 42

    4.3 Results 43

        4.3.1 Clustering of AL neuron activity patterns based on spiking and sub-threshold information. 43

        4.3.2 Clustering of AL neuron activity patterns based on spiking information alone. 46

        4.3.3 Distribution of PNs and LNs in the different cluster trees. 48

    4.4 Discussion 49

        4.4.1 Holistic or simplistic - how much information is necessary to distinguish meaningful clusters? 50

        4.4.2 Science or fiction - may electro-physiological characteristics be used to predict AL-neuron morphology? 51

        4.4.3 Utile or futile - why do we need established electro-physiological groups of AL neurons? 52

5 Conclusion and Outlook. 53

Summary 59

Zusammenfassung 63

Danksagung 67

Bibliography 69

CHAPTER 1

General Introduction.

"No sooner had the warm liquid, and the crumbs with it, touched my palate than a shudder ran through my whole body, and I stopped, intent upon the extraordinary changes that were taking place within me. An exquisite pleasure had pervaded me, unconnected with anything, with no suggestion of its origin."

Clearly, the first person narrator of Proust's "Remembrance of Things Past" who contemplates the exquisite pleasure of food scent, is human - not a dog, a rat, a bee, or any other species that relies heavily on its olfactory skills to manage every day life. These animals would neither have to muse about the cause of the exquisite pleasure, nor would they have had to search for the identity of the eliciting odour over a time long enough to write a three pages excerpt on it. The importance of olfactory cues in triggering appropriate behaviour makes solid performance in recognition and discrimination of meaningful odours a necessity to most animals.

Signal processing in olfactory brain areas of mammals and insects - a one for all solution.

Before recognition or discrimination is possible, the organic chemicals that constitute an odour, must be transformed into a neural signal that can be interpreted by the brain. The process of stimulus encoding is common to all sensory systems. It is accomplished by receptors sensitive to stimuli of only one modality and neurons in those brain areas that are dedicated to the same modality. In the mammalian olfactory system, olfactory receptor neurons (ORNs) project directly in the olfactory bulb (OB). Relay-neurons, the mitral/tufted cells (M/T cells), send information onwards to upstream, multi-modal brain areas. As an exemplary case of convergent evolution, the olfactory system of the insect largely resembles the situation in the mammal (Hildebrand and Shepherd 1997): ORNs on the antennae, provide input to the antennal lobe (AL), which homologue to the OB. AL relay-neurons, termed projection neurons (PNs) carry the AL's output to upstream, multi-modal brain areas. As a consequence of its seemingly simple construction, the olfactory system is considered a flat processing stream (Wilson and Mainen 2006). While this might be true compared to e.g. the visual system that, depending on the species, can encompass up to 40 dedicated areas (van Essen 2003), the olfactory system still exhibits a remarkable computational power.
This becomes apparent, in that odour coding is not simply matching stimuli to receptors and passing on the signal. The olfactory system has to provide the upstream cognitive areas with neither more nor less than all information that is relevant to the organism(for review see Gottfried 2010). This

includes odour identity, odour concentration, the separation of a relevant stimulus from background odour-environment, the amalgamation of many stimuli into an odour-mixture, as well as temporal properties of the stimulus. Figuratively, the same task has to be accomplished in the same manner by every other sensory system, but as noted above it gets distributed throughout an extensive processing path with many different areas. How is the olfactory system to achieve the same task with its limited resources?

Multiple mechanisms, like signal amplification, gain control, contrast enhancement, odour specific latencies, and firing synchrony are evident to aid odour coding. Most of these processes are thought to happen in the OB/AL. In order to understand how they might be implemented, we will have to know a little more about the architecture of these olfactory brain areas. First, I will give a briefly describe the suggested implementation of processing mechanisms in the OB, to subsequently give a more detailed description of the AL.

Olfactory processing in the OB - an example for the functionality of interneuron sub-populations!

The OB consist of multiple functional subunits with high synaptic density, the glomeruli. It is here that ORNs and M/T cells meet. The remarkably high convergence of many, spatially distributed ORNs relating to few M/T cells is a means to amplify the signal and minimize noise as a consequence of local fluctuation of stimulus concentration (Laurent 1999). However, most of the described mechanisms cannot be accomplished by ORNs and M/T cells alone but are mediated by interneurons, which constrain their neurites to the OB. Inhibitory periglomerular cells, excitatory external tufted cells, and excitatory short axon cells form synapses in the glomerular layer of the OB. A second layer is constructed by synaptic circuits beyond the glomeruli, where inhibitory granule cells and branches of M/T cells establish inter-glomerular interaction (for reviews sees Kay and Stopfer 2006; Wilson and Mainen 2006).

Most of the periglomerular cells synapse within single glomeruli, where they act as recurrent, presynaptic inhibitors of ORN activity , as well as inhibitors of M/T cell activity. These properties make them candidate neurons to realise gain control, keeping the OB levelled within its dynamic range. The principal connectors between glomeruli are short axon cells (Aungst et al. 2003). These interneurons excite perglomerular cells and external tufted cells in distant glomeruli. Short axon cells are suggested to contribute to both, gain control (Linster and Cleland 2009) and contrast enhancement (Aungst et al. 2003; Hayar et al. 2004). Granule cells form synapses with secondary dendrites of M/T cells outside the glomerular layer. They interconnect M/T cells that reside in same and different glomerli and are a peculiarity of the nervous system in that they are axonless. Granule cells perform complex dendrodendritic interactions, which include reciprocal synapses and auto-inhibition, and are thought to engage in multiple processing tasks (for review see Shepherd et al. 2007). Widespread lateral inhibition that these cells mediate is suggested to serve contrast enhancement (Mori and Shepherd 1994) and perhaps even columnar organisation of the OB (Willhite et al. 2006). Local inhibition of M/T cells within the same glomerulus might be the cause of oscillatory synchrony (Rall and Shepherd 1968). Already this coarse simplification of what is likely to happen in the OB, clearly states that interneurons of the mammalian olfactory system can be grouped in sub-populations that can be attributed different functions and decisively influence processing of odour.

Olfactory processing in the AL - an example for the functionality of interneuron sub-populations?

To what extend does the outlined division of labour in networks of functionally different interneurons transfer to the homologue insect AL? The AL is, just like the OB, constituted of multiple glomeruli, in which many ORN axons converge with dendrites of few PNs. Insect PNs are found abundantly in two variations: uni-glomerular, i.e. innervating only one glomerulus, or multi-glomerular, i.e. innervating several glomerlui (Homberg et al. 1988; Malun et al. 1993; Mobbs 1982; Stocker et al. 1990). Unlike the OB, the AL does not have a second layer of interaction. Even though uniglomerular PNs have been shown to extend secondary branches to the outside of their innervated glomerulus (Müller et al. 2002), where some few synapses are found (Gascuel and Masson 1991), the vast majority of computing takes place within the glomeruli.

Local interneurons (LNs), with neurites restricted to the AL, establish both, inter-glomerular and intra-glomerular connections. Within and between species of insects, various LN morphologies have been described (Chou et al. 2010; Christensen et al. 1993; Dacks et al. 2010; Flanagan and Mercer 1989; Fonta et al. 1993; Seki and Kanzaki 2008; Seki et al. 2010; Stocker et al. 1990). These are commonly differentiated according to the density of arborisations within single glomeruli and symmetry of innervations between different glomeruli. Whether any of the possible morphological sub-populations in one species relate to that of another is uncertain. Approaching the question of LN sub-populations from a mechanistic point of view, the situation is equally complex. Measurements of ORN and PN activity under conditions of selective suppression of input by lesions, or competitive input in the form of odour mixtures, provide evidence for LN mediated gain control, in both honeybee (Deisig et al. 2006) and drosophila (Olsen et al. 2010; Olsen and Wilson 2008; Silbering and Galizia 2007). However, there is no morphologically identified sub-populations of LNs explicitly connected with gain control in the AL. LNs are generally thought to be inhibitory, but excitatory populations are postulated to exist in the honey bee (Malaka et al. 1995) and shown in drosophila (Shang et al. 2007). Unlike short axon cells in the OB, the cholinergic population of LNs in drosophila seems to implement a mechanism of lateral excitation that broadens PN response spectra. Even though antibody stainings prove the existence of cholinergic LNs, and physiological recordings elucidate their functionality, evidence for a specific cell morphology is lacking. The issue of contrast enhancement by lateral inhibition is controversially debated (Galizia 2008; Wilson and Mainen 2006). Experimental and computational evidence from the honey bee suggest that lateral inhibition might follow a map of functional similarity and be mediated by non-GABAergic LNs (Deisig et al. 2010; Linster et al. 2005; Sachse and Galizia 2002; Sachse et al. 2006). Fast oscillations have repeatedly been discussed as being generated by GABAergic LNs (Christensen et al. 1998a; Stopfer et al. 1997; Wilson and Laurent 2005). Alteration of spatial activity patterns in the honey bee AL predicts the corresponding LNs to innervate many glomeruli homogeneously (Sachse and Galizia 2002). Direct evidence in support of this notion was only recently provided by a study in drosophila (Tanaka et al. 2009). Interestingly, GABA has likewise been associated with LN mediated inhibition in the context of gain control, temporal patterning, and lateral inhibition. The different mechanisms are suggested to arise as a consequence of morphologically distinct LNs (Tanaka et al. 2009), action on two types of receptors at the receiving synapse (Wilson and Laurent 2005), or differential activation of a single type of LN (Christensen et al. 2001, 1998b).

Taken together, LNs of the AL differ in many properties, and seem to be involved in as many processing mechanisms as interneurons of the OB. Given the diversity of LNs within and between species, together with the difficulty of assigning these specific functions, we end up with two possible, extreme scenarios: either the AL is endued with multiple functional AL sub-populations that implement related but not identical processing mechanisms(Yuste 2005), or the different properties of LNs are a continuum of random variations of one basic type of LN that serves the realisation of multiple mechanisms (Parra et al. 1998). While the truth most probably lies somewhere in-between, the question will only be resolved by accumulation of direct evidence from LNs.

Objectives of this work.

In the work at hand I adopted the view, that functional LN sub-populations do exist and help to implement important processing mechanisms in the AL. In order to trace down characteristics by that a) functional LN sub-populations might be unambiguously identified, and b) LNs differ from PNs, I approached the problem with different methods.

In chapter two, I present the results of combined physiological and morphological investigation of AL neurons. I used dynamic odour stimuli, such as might be experienced in nature, to find groups of neurons that differed in both coding strategy and response latency. By subsequent staining, I could illuminate whether neurons following different coding strategies are also morphological distinct.

In chapter three, I exemplify morphological properties of LNs. My first interest here was to find out how LN morphology relates meaningfully to structures of AL input and output. Further, I was asking if a description of LN morphology including more detailed criteria than those commonly used for LNs of the honey bee, would still allow to establish groups of distinct phenotypes.

In chapter four, my guiding question was, in how far AL neurons can be classified by means of their odour-evoked spiking patterns. For this purpose, I used descriptive values for electrophysiological properties of spiking neurons, based on which I performed PCA and hierarchical clustering.

CHAPTER 2

Elemental and configural odour-coding by antennal lobe neurons of the honey bee.

Contents

2.1	**Introduction**	5
2.2	**Materials and Methods**	7
	2.2.1 Animal preparation	7
	2.2.2 Odour stimulation	8
	2.2.3 Electrophysiology	8
	2.2.4 Morphology	9
	2.2.5 Data analysis	10
2.3	**Results**	11
	2.3.1 Odour concentration can be used to enhance stimulus specific latency shifts in AL neurons.	11
	2.3.2 AL-neurons differ in their response patterns	12
	2.3.3 Elemental and configural coding both occur in AL neurons.	13
	2.3.4 AL-neurons are active sequentially.	16
	2.3.5 Hetero LNs are involved in configural as well as elemental processing.	17
2.4	**Discussion**	18
	2.4.1 Broad and narrow tuned LNs - functional subgroups?	19
	2.4.2 Puzzles of suppression and excitation - a complex AL blueprint?	19
	2.4.3 Hetero LNs - multi-function neurons?	20

2.1 Introduction

The insect olfactory system has become an important model system to investigate not only olfactory but also general neural processing mechanisms (Hildebrand 1995; Sato and Touhara 2009). The honeybee's ability to differentiate between many odors makes her a first rate organism to study olfactory coding (Menzel et al. 1996).
Olfactory coding starts at the antennae where odour molecules are detected. Information is then transferred to the antennal lobe (AL), the insect's primary olfactory neuropile. The AL is thought to

reformat the input signal before it is relayed onwards to higher processing areas (Galizia 2008; Kay and Stopfer 2006).

The AL consists of approximately 160 functional subunits with high synaptic density, the glomeruli. In each glomerulus three classes of neurons synapse onto each other: olfactory receptor neurons (ORN), projection neurons (PNs), and local interneurons (LNs). ORNs detect odor molecules at the antenna and form the input level of the AL. Each glomerulus receives sensory input from one type of ORNs. ORNs branch throughout the superficial layer of the glomerulus, its "cap". PNs send their axons from the AL to higher processing areas and establish the output level of the AL. Some of these neurons invade several glomeruli (multiglomerular PNs), others only a single glomerulus (uniglomerular PNs) (Fonta et al. 1993). LNs branch exclusively within the AL and interconnect glomeruli. Moreover they can interconnect the cap and the central "core" within one glomerulus.

Input level, that is ORN signals, as well as output level, that is PN signals, have been studied intensively in various species of insects unravelling several AL mechanisms such as contrast enhancement (Linster et al. 2005; Sachse and Galizia 2002), gain control (Deisig et al. 2006; Olsen et al. 2010; Olsen and Wilson 2008), and firing synchrony (Lei et al. 2002; Perez-Orive et al. 2002; Tanaka et al. 2009). LNs are the likely mediators of these mechanisms, but in comparison with ORNs and PNs they have received less attention. In accordance with the multiple tasks LNs get associated with, a broad variety of LNs have been described (Chou et al. 2010; Christensen et al. 1993; Dacks et al. 2010; Flanagan and Mercer 1989; Fonta et al. 1993; Seki and Kanzaki 2008; Seki et al. 2010; Stocker et al. 1990). In hymenoptera, the honeybee in particular, two main morphological groups are distinguished: homo LNs uniformly innervate many glomeruli, and hetero LNs innervate one glomerus densely and several sparsely. These groups however can be further differentiated both in terms of morphology (Dacks et al. 2010; Flanagan and Mercer 1989), as well as histochemistry (Dacks et al. 2010; Kreissl et al. 2010; Nässel and Homberg 2006; Schäfer and Bicker 1986). Physiological recordings from single honeybee LNs did not allow for functional grouping so far. Responses of hetero LNs often correspond to the PN response profile of their densely innervated glomerulus (Galizia and Kimmerle 2004), but more often LNs are reported to respond rather unspecifically to most olfactory stimuli (Abel 1997; Krofczik et al. 2009; Sun et al. 1993). Most LNs respond to stimulation with excitation, but inhibition as well as complex response patterns composed from intervals of both, inhibition and excitation, have been described (Flanagan and Mercer 1989; Sun et al. 1993). Like in the moth, LN response latencies in the honeybee were found to be shorter than those of PNs suggesting that signal transfer from ORNs to PNs is mediated via LNs (Christensen et al. 1993; Krofczik et al. 2009). Furthermore, morphological evidence to the existence of reciprocal synapses (Gascuel and Masson 1991) opens up for the possibility of a complex synaptic layout including both, LN mediated and direct ORN-PN signal transduction.

In the present study we conducted intracellular recordings and morphological reconstructions from single AL neurons (LNs, as well as PNs) of the honeybee *Apis mellifera*. Our main objective was to learn how individual neurons encode complex information content.

A common procedure to find a cell's receptive field is by systematically screening a broad variety of potential stimuli. Considering the uncountable number of volatile compounds that are potential odors this approach seems not to be applicable in olfaction. Instead, we decided to offer a small set of stimuli that covered a range of stimulus aspects in the way they occur in nature, i.e. quality, quantity, and temporal complexity in a mixture. The sum of these aspects a constitutes stimulus' identity.

Odour quality (Krofczik et al. 2009) as well as concentration (Christensen et al. 1993; Stopfer et al. 2003) impact the response onset measured in electro antennogramm recordings (EAG) as well as the latency of individual PNs. The resulting latency code reliably predicts stimulus identity (Junek et al. 2010). We made use of this phenomenon and designed a paradigm in which response latency served as an indicator of the effective stimulus' identity. We chose two monomolecular odors (1-Octanol and 2-Heptanone) that naturally occur in the honeybee's environment as both, components in floral mixtures (Baraldi et al. 1999; Omata et al. 1990; Tollsten and Knudsen 1992) and pheromones (Balderrama et al. 1996, 2002). We presented the single components alone, their temporally perfect mixture (both components with synchronized odor onset), and their temporally imperfect mixtures (both components with odor onsets shifted in respect to each other). In doing so, we created a reproducible version of a natural, dynamic odor environment. We expected to find neurons to code our stimuli in one of two possible ways: configural, in which case the neuron responded to the mixture as to a new odor, or elemental, with the mixture response corresponding to the response to one of the components. Based on the assumption that shifts in response latency within a cell are specifically correlated with stimulus identity, we compared absolute latencies between stimulus conditions to distinguish configural and elemental coding strategies.

2.2 Materials and Methods

2.2.1 Animal preparation

Worker honey bees (*Apis mellifera*) were caught at the entrance of the hive or at a feeder, immobilized by cooling, and mounted in custom made plexiglas holders. The bees were allowed to acclimate to the new environment for 1-6 hours before the experiment started.

Antennae were immobilized with Eicosane (melting point 37 C°; Sigma-Aldrich Chemie GmbH, Germany) whilst head and mandibles were immobilized with Deiberit 502 (melting point 60 C°; Boehme-Schoeps, Germany). To reduce brain movements the esophagus was detached from its muscles. The head capsule was opened between the median ocellus and the base of the antennae. Glands and tracheal sheaths were removed carefully. The exposed brain was kept moist during recordings by dribbling saline onto it if necessary (in mmol $^{-1}$: 130 NaCl, 6 KCl, 4 $MgCl_2$, 5 $CaCl_2$, 10 Hepes, 25 D-Glucose, 160 sucrose, pH 6.7, 500 mosml).

8 Materials and Methods

2.2.2 Odour stimulation

During recordings bees were placed at a fixed distance to the stimulation device, a custom build olfactometer similar to a previous published model (Galizia et al. 1997). Stimulus delivery was controlled by TTL pulses triggered by the recording software, Clampex (AxonInstruments Inc., USA). Relative gas flow through individual channels of the olfactometer was measured at its exit with a photoionization detector (Aurora Scientific Inc., Canada). The primary odorants 1-Octanol and 2-Heptanone, diluted in mineral oil were used for stimulation. Odorant concentration was $0.5*10^{-2}$ for 1-Octanol and $1*10^{-2}$ for 2-Heptanone. Concentration specific effects were checked for by recording EAGs with these same dilutions ($0.5*10^{-2}$ 1-Octanol, $1*10^{-2}$ 2-Heptanone), reversed dilutions ($1*10^{-2}$ 1-Octanol, $0.5*10^{-2}$ 2-Heptanone) and equal dilutions ($1*10^{-2}$ 1-Octanol, $1*10^{-2}$ 2-Heptanone). Airborne stimuli were delivered in a constant stream of clean air (1.2 m/sec) that was directed to both antennae. Stimulus duration was 800 ms at an inter-trial interval of 1800 ms. Three trials using identical stimuli followed each other in immediate succession constituting one stimulus block. An interval of 5000 ms separated two blocks from each other (Fig.2.1). A completed recording consisted of the presentation of five stimulus blocks and two control blocks: each of the two primary odorants, their temporally perfect mixture, their two temporally imperfect mixtures, a control of mineral oil and a control of pure air. In an imperfect mixture the onset of the first odorant was delayed by 50 ms in respect to the second odorant. The sequence between presentations of different blocks was pseudo-randomized.

2.2.3 Electrophysiology

Three types of electro-physiological experiments were performed: Electro antennogram (EAG) recordings, intracellular recordings from single AL neurons, and parallel recordings of both of these. EAG recordings served to validate the stimulus apparatus and to analyse odorant concentration-specific effects in ORNs. Intracellular recordings were used to investigate the role of single AL neurons within the AL network. Parallel recordings with both methods allowed determination of the response onset of single AL neurons with respect to a precise reference.

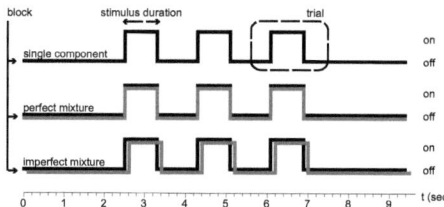

Figure 2.1: Stimulus protocol. Exemplified for an imperfect mixture. Stimulus duration was 800 ms, followed by an 1000 ms interval. Each stimulus repetition was regarded as one trial. One block consisted of three trials with identical stimulation. The onset delay for odorants in imperfect mixture blocks was 50 ms.

Glass electrodes were pulled from borosilicate capillaries (GC150F-10, Clark electronic instruments, UK) by means of a horizontal Puller (P-97, Sutter Instruments Co.,USA). Sharp electrodes (100–250 MΩ) used for intracellular recordings were tip-filled with fixable fluorescent dye (4% Alexa 488 hydrazid in 0.2 M KCl, 4% Micro Ruby in 0.2 M KAcetate or 3% Lucifer Yellow in 0.1% LiCl). Blunt electrodes (5–20 MΩ) used for EAG recordings were filled with 0.2 M NaCl.

The sharp electrodes were placed on the AL and gradually advanced employing a micro-manipulator (Kleindiek Nanotechnik, Germany) until a cell was impaled. For EAG recordings, a blunt electrode was placed on the antenna using a second micro-manipulator (Brinkmann Instrumentenbau, Germany). In cases where both signals were recorded in parallel, the EAG was always taken from the antenna ipsilateral to the AL recorded from. A common reference electrode was placed through a small incision between the lower ocelli.

Recordings were performed in current-clamp mode, using an Axoclamp 2B Amplifier (gain 10, AxonInstruments Inc.,USA). EAG signals were additionally amplified by means of a custom-build external amplifier (gain 10). A 50 Hz filter (Hum-Bug, Quest Scientific, Canada) removed line hum. Data were digitized using the Axon Interface, DigiData 1200B (AxonInstruments Inc., USA) and stored on hard-drive using Clampex 8.2 (AxonInstruments Inc.,USA).

2.2.4 Morphology

After an intracellular recording was finished successfully, the dye loaded in the tip of the sharp electrode was iontophoretically expelled, with the polarity of the current pulses (0.2 s width, 2 Hz, 1-4 nA) chosen according to the dye's charge.

Subsequently, the sensory tracts of the penetrated AL were labelled with neurobiotin (Vector Laboratories Inc., USA). For this purpose, the cuticle previously removed from the head capsule was now replaced and closed carefully again using Eicosan. The ipsilateral antenna was brought in an upright position and surrounded by a basin made from Vaseline that was filled with 2% neurobiotin (in aq.dest.). The antenna was cut at the scapus and the neurobiotin was given 2-3 hours to be taken up by the antennal nerve stump.

For morphological preparations, brains were removed and fixated in 4% para-formaldehyde for 3 hours at RT, or over night at 4°C. Subsequently, preparations were washed 3 times for 10, 30, and 45 minutes in phosphate buffered solution and incubated in 0.5% avidin-coupled fluorescent dye to visualise the neurobiotin in the OSNs (either AMCA-avidin, or cy3, depending on the single-cell marker) for at least 5 hours. Brains were then washed again 3 times for 15, 30 and 45 minutes, dehydrated in an ascending ethanol series, degreased for 20 min in Xylol and finally embedded into DPX mounting medium (Fluka, Sigma-Aldrich Chemie GmbH). To visualise staining results stacks of images were taken with a Zeiss LSM 510 Meta confocal microscope (Carl Zeiss AG, Germany).

2.2.5 Data analysis

EAG recordings were filtered off-line (100 Hz lowpass cutoff) and analysed using custom written routines in R (http://www.R-project.org). Likewise, sharp recordings were filtered off-line (10 kHz lowpass cutoff). Spike were detected, using custom written routines based on the open source R packages SpikeOMatic (Pouzat et al. 2004) and STAR (Pouzat and Chaffiol 2009). To determine firing rate and response latencies, algorithms provided by the open source Matlab toolbox FIND (http://find.bccn.uni-freiburg.de) were employed. Analysis of single cell data was chosen so as to maximize comparability to related work (Krofczik et al. 2009). Image processing of confocal stacks and reconstruction of cell morphology were achieved using AMIRA 5.1 software (Mercury Computer Systems, Germany).

Temporal EAG analysis

EAG recordings were averaged over repeated trials and low-pass filtered with a cut-off frequency at 100 Hz. Response onset was defined as the relative maximum preceding the steepest negative slope of the potential drop which demarcated an odour-response. This point was found to be least affected by temporal displacement attributable to response amplitude and, hence, evaluated as most reliable.

Response latency analysis

Absolute latency, that is the mean latency across trials, and relative latencies, that is trial-to-trial differences in latency, were calculated with one of three methods (1-3). The method was chosen based on the respective firing pattern.

1) Latencies of cells that responded to stimulation with an increased firing-rate were estimated based on the derivative of the trial-aligned firing rate as described elsewhere (Meier et al. 2008). This method processes the data in four successive steps. First, the derivative of each single trial spike train of a given cell was estimated by convolving with an asymmetric Savitzky-Golay filter (Savitzky and Golay 1964) (polynomal order 2, 300 ms width, Welch windowed). Second, all single trial-derivatives were optimally aligned, finding the greatest possible pair-wise cross correlation (Nawrot et al. 2003). The resulting time-shifts correspond to each trial's relative latency. Their standard deviation σ gives a measure for the across-trial latency variability. Third, the single trial spike trains are temporally aligned by shifting each by its individual relative latency. Fourth, the aligned spike trains were merged into one train, representing the cells activity pooled over trials. The convolution of this merged spike train with the same asymmetric Savitsky-Golay filter gave an estimate about the derivative of the cell's absolute firing rate, based on which the absolute latency within a given block of stimulation was determined. The stimulus specific absolute latency was defined as that point in time where the slope of the firing rate is steepest, that is the derivative's maximum.

2) Latencies of cells that responded to stimulation with a decrease in firing rate were estimated with an approach nearly identical to 1), but instead of the steepest rising slope, the steepest falling slope of the absolute firing rate was defined as response onset.

2.3. Results

3) Latencies of cells that had very low spontaneous activity and responded to stimulation with a membrane depolarisation ridden by one or few single spikes were estimated based on spike peak time rather than rate. The membrane depolarisation in these cells was taken as indicative for an apparent response. The response latency was defined as the peak time of the first spike riding such a depolarisation.

Single cell morphology

Confocal image stacks were processed using Amira 5.1 software (Visage Imaging GmbH, Germany). Neurons were reconstructed using the filament editor without further estimation of neurite diameter. Location and size of single glomeruli were registered by interactive segmentation based on OSN mass fills. Glomerulus identity was determined by visual inspection and comparison with the morphological atlas of the honey bee (Galizia et al. 1999, http://neuro.uni-konstanz.de/honeybeeALatlas). To compare innervation patterns with spacial patterns of AL activity in response to the stimuli applied, the physiological atlas of the honey bee (http://neuro.uni-konstanz.de) was consulted.

2.3 Results

2.3.1 Odour concentration can be used to enhance stimulus specific latency shifts in AL neurons.

Odour quality as well as odour concentration are aspects of stimuli that influence neural response onset. For our odour-mixture experiments we needed two stimuli with a distinct response delay. For this purpose we chose two different odorants and increased difference in latency even further by applying them in different concentrations.

We recorded EAGs and found that irrespective of odour quality the latency of the higher concentrated odorant was significantly shorter when compared to the lower concentrated odorant (two-way anova, $F \langle 0.000$; Tukey test, $p \langle 0.000$; mean difference = 22.646 ms). Accordingly, odour concentration can be used to create stimuli with distinct, characteristic latency shifts.

In order to investigate how ORN latency shift transfers to the AL network we conducted simultaneous intracellular and EAG recordings (Fig.2.2 A). The odorant at lower concentration evoked a smaller EAG amplitude (Fig.2.2 B) as well as a weaker firing rate in phasic-tonic AL neurons (Fig.2.2 C). As was the case in the EAG, the latency in AL neurons differed significantly between the two stimuli with the higher concentrated stimulus evoking shorter response latencies (one-sided, paired t-test, $p = 0.048$). Indeed, the difference between mean response onsets of EAGs and AL neurons was identical for both stimuli (37 ms; Fig.2.2 D). Thus, concentration dependent latency shifts originate in ORN activity and transfer directly to AL neurons.

Simultaneous recording of AL neurons and EAG further allowed us to report the response latency of single neurons more accurately. Timing of olfactory stimuli is always constrained by laboratory

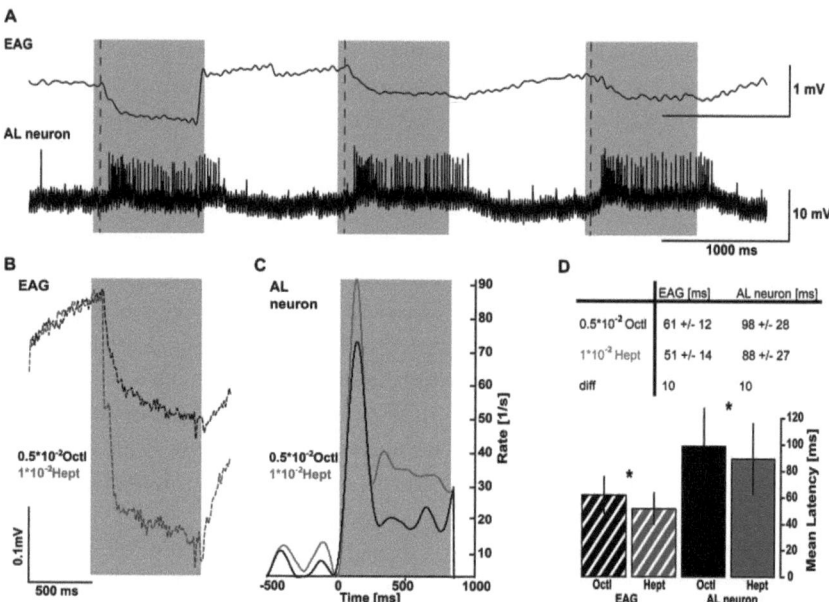

Figure 2.2: AL neurons inherit a concentration dependent latency shift from peripheral neurons. (**A**) Exemplary traces of simultaneous EAG and single cell recording. Light grey background indicates stimulus delivery time. Dotted lines indicate the time of stimulus detection on the antennae, measured by EAG (upper trace). (**B**) EAG amplitude differs for different stimuli. Traces shown are averaged (n= 11). (**C**) AL neurons mean firing rate differs for different stimuli. Rate functions are aligned and averaged over all neurons which responded with a phasic-tonic pattern (n= 12). (**D**) Response latencies of EAG and AL neuron recordings differ significantly for different stimuli, but keep the same relative time shift (37 ms).

conditions, like air turbulences or distance between odour source and receiver (Vetter et al. 2006). These constraints are difficult to standardize between different laboratories. Comparability of data between laboratories is increased when latencies are estimated with respect to a reference derived from the animal rather than the stimulation machinery. The EAG reliably reflects stimulus arrival at the antennae. Hence we used the mean EAG response onset as a reference time point to estimate single AL neuron latencies.

2.3.2 AL-neurons differ in their response patterns

We found three different types of AL neuron responses: phasic-tonic excitation, inhibition, or few spikes riding on a depolarisation (Fig.2.3 A). The most abundant type of response was a phasic-tonic excitation (n = 12). Fewer neurons responded with inhibition (n = 4) or a strong membrane depolarisation which was ridden by one or few, exactly timed spikes (n = 5).

In order to obtain a reasonable estimate of an individual cell's latency we defined the onset criterion with respect to the cell's response pattern (Fig.2.3 B). Responses were highly reliable in all probed neurons. A stimulus that elicited a response in a given neuron once, did so in every trial.

Furthermore, response type (excitation, inhibition, single spike) was equal within each neuron for the stimuli used in this experiments. Cells with different response types did not differ systematically in their mean latency (Anova, F = 2.265, P = 0.133). This finding, together with previous reports (Sun et al. 1993) about cells which exhibited excitation as well as inhibition, suggests that functional subgroups of neurons can not be established based on their response type alone.

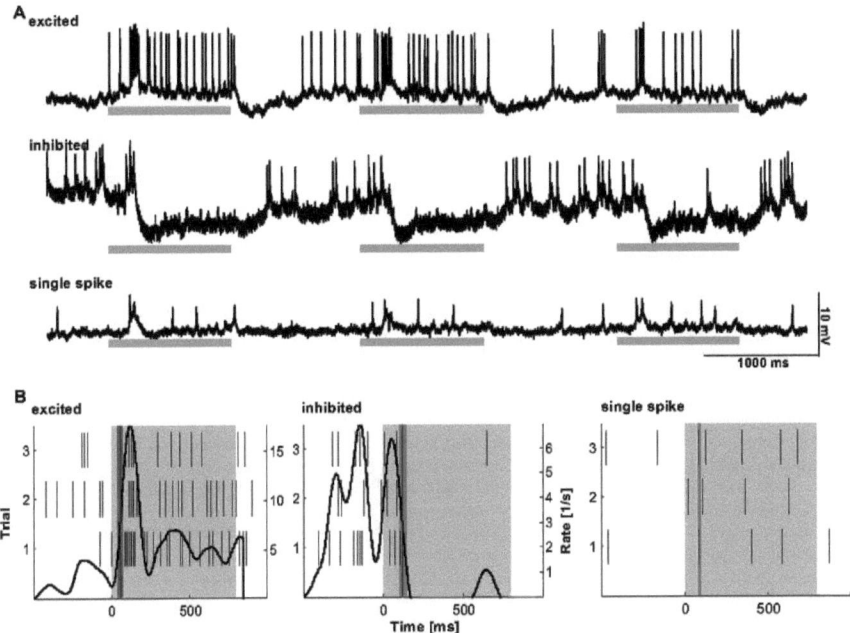

Figure 2.3: **Response types of intracellularly recorded AL neurons.** (A) Response patterns of AL neurons differed in detail but fell into three types: phasic-tonic excitation (top), inhibition (middle), or few spikes on a depolarisation (bottom). (B) Estimation of response latency. Spiking activities shown for three repeated trials under identical stimulation. Superimposed traces indicate the response rate function. Vertical lines mark the estimated response onset. Surrounding grey bars indicate their across trial variability. Response onset for phasic-tonic excitation is defined as the point of steepest rising in the rate function after stimulus onset (left). Response onset for inhibition is defined as the point of steepest falling (middle). Response onset for neurons with continuously sparse firing is not well captured by a rate function and best characterized by the peak time of a single spike after depolarisation onset (right). Light grey background indicates stimulus delivery time.

2.3.3 Elemental and configural coding both occur in AL neurons.

Our experimental design was aimed at creating a dynamic odour environment, such as might be experienced in nature, but with well controlled features to allow for standardized experiments. To this end, we presented single odorants and their temporally perfect and imperfect mixtures. We asked

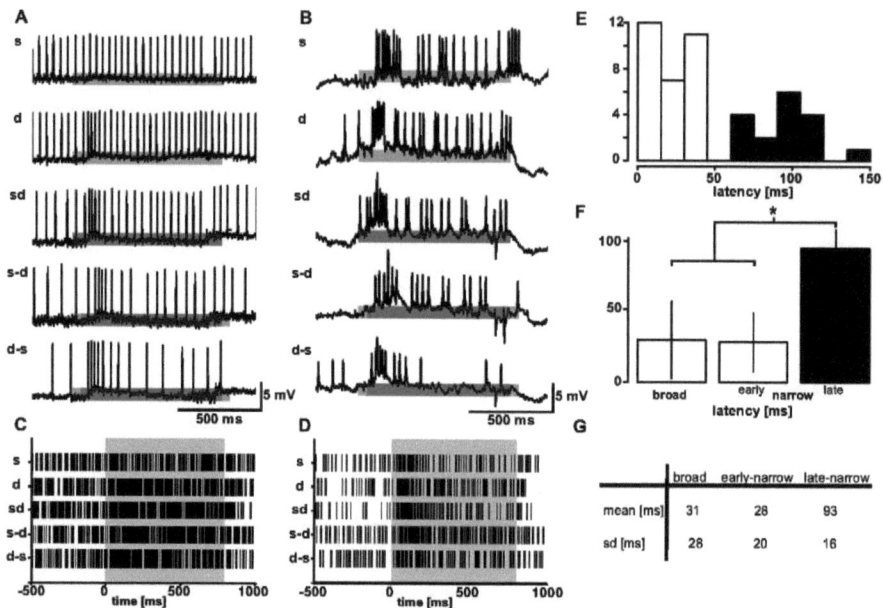

Figure 2.4: Grouping 'elemental neurons', according to tuning and mean cell latency (A) Example of a narrowly tuned elemental neuron. The dominant odorant (d) evokes responses both as single stimulus and in mixtures (sd = perfect mixture, s-d/d-s = imperfect mixtures). The subordinate odorant (s) does not contribute to the response. Grey bars indicate the time throughout which valves are open. (B) Summed spike trains comprising all narrowly tuned elemental neurons (n = 5). Note the lack of response to the subordinate odorant (s), as compared to the other conditions. (C) Example of a broadly tuned elemental neuron. Both odorants evoke responses, but in imperfect mixtures a latency shift indicates preference of odorant d. (D) Summed spike trains comprising all broadly tuned elemental neurons (n = 7). Note that subordinate (s) as well as dominant (d) odorant evoke a response. (E) Irrespective of tuning, neurons' latencies cluster into an early (white) and a late (black) group. (F) Based on the combination of latency and tuning, three neuron groups get apparent. Broadly tuned neurons always respond early (white, left), narrowly tuned neurons distribute in an early (white, middle) and a late group (black), with significant different latencies (paired t-test, one-sided, p = 0.001). (G) Table of mean cell latencies sorted according to group.

whether it was possible to find groups of neurons with similar mean response latencies but that follow different coding strategies, indicating the existence of functional subgroups.

About half of all successfully recorded AL neurons responded to one of the single components even within a mixture (12 of 21, 'elemental neurons'). The remaining cells (n = 9, 'configural neurons') responded to the mixtures as to a new odour.

Based on their tuning-properties, elemental neurons formed two subgroups. The first group (n = 7) was narrowly tuned in that it responded only to one of the single components, its dominant (d) odorant. Response latency to the perfect mixture, and the imperfect mixture in which odorant d was presented first, corresponded to the response latency to odorant d alone. Responses to the imperfect mixture in which odorant d was presented as the second component were shifted by about 50 ms, corresponding to the delay between odour pulses. Hence we conclude that these neurons responded to odorant d alone, and were not influenced by the presence of odorant s (Fig.2.4 A,B). The second group of elemental neurons was broadly tuned (n = 5). These neurons responded to both of the single components. However, in the mixture only one odour (d) contributed to the response, while the other (s) did not have any apparent impact (Fig.2.4 C,D). Based on their latencies, elemental neurons formed two distinct clusters, one of short latencies below 60 ms and another of long latencies above 60 ms (Fig.2.4 E). All broadly tuned neurons fell into the short latency cluster, while narrowly tuned neurons distributed into both clusters. This resulted in three elemental neuron subgroups: broadly tuned with short latencies (n = 5), narrowly tuned with short latencies (n = 4) and narrowly tuned with long latencies (n = 3) (Fig.2.4 F,G).

Responses of configural neurons were more diverse, both in terms of tuning and latency. One cell responded only to temporally perfect stimuli (Fig.2.5 A), some to mixtures only (n = 2), but most cells responded to mixtures as well as single compounds (n = 6, Fig.2.5 B). Latencies of configural neurons scattered broadly around 60 ms and thus concentrated exactly between the groups of elemental neurons with short- and long-latencies (Fig.2.5 C,D). Unlike for elemental neurons latencies within one configural neuron could be short for one stimulus block and long for another one. As a general trend fastest responses were evoked by single compounds (62±26 ms) and lowest by temporally imperfect mixtures (84±49 ms, paired t-test, p⟨0.05, Fig.2.5E).

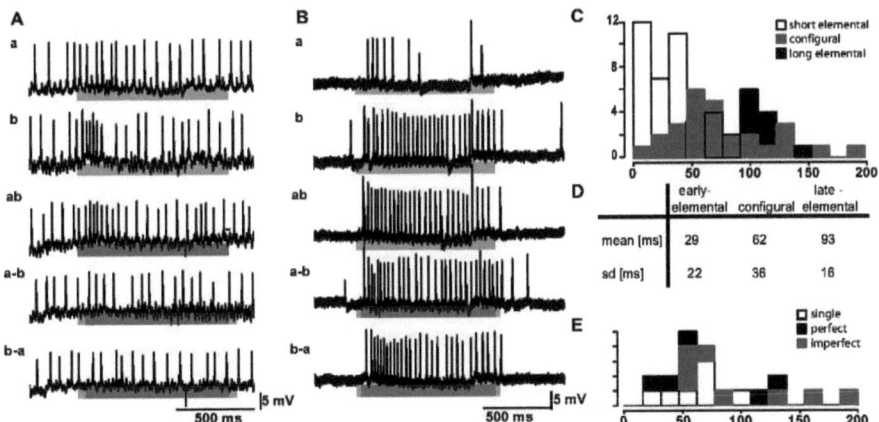

Figure 2.5: 'Configural neurons' and distribution of latencies in respect to 'elemental neurons'. (A) Responses to temporally perfect stimuli only (Latencies [ms]: b 94, ab 132). Grey shades indicate stimulus delivery time (B) Example of an AL neuron which responded to single compounds and coded mixtures configuraly (n = 6, Latencies of the individual neuron [ms]: a 62, b 49, ab 61, a-b 67, b-a 87). (C) Latencies of neurons with configural responses (grey) concentrate between the early (white) and a late (black) group of neurons with elemental responses. (D) Table of mean cell latencies sorted according to group. (E) In neurons with configural responses, single compounds (white) evoke shorter latencies than imperfect mixtures (grey).

2.3.4 AL-neurons are active sequentially.

We compared population rate functions between subgroups in more detail (Fig.2.6). Early-latency responses of 'elemental neurons' were either excitations, or single spike. We pooled these and estimated a common rate function. Late-latency responses of 'elemental neurons' were either excitations or inhibitions, which we separated in two rate functions. Amongst the configural neurons all types of responses were represented. We pooled all non-inhibited responses in a common rate function.

Superimposition of these rate functions illustrates that activity peaks of the three groups follow each other in immediate succession. Rate functions of elemental neurons show a rapid rise and a distinct peak, while that of configural neurons develops rather gradually into a sustained plateau, reflecting the scatter of latencies within each single cell. This large scatter of latencies within each configural neuron makes it difficult to place it somewhere i a fixed processing circuitry. We hypothesise that that these neurons are in fact, in dependence of stimulus context employed in different circuits. Early responding elemental AL neurons (white) precede the positive rate peak of excited (black; Fig.2.6 A), as well as the negative rate peak of inhibited late responding elemental neurons (black; Fig.2.6 B). Interestingly, the rate function of inhibited late responding neurons appears to have a small activity boost, which relapses just before early responding neurons reach their maximal response frequency. This suggests that early-responding AL-Neuron impact late-responding AL neurons in at least two different ways, delayed information transmission and inhibition. Honey bee LNs have been shown to respond faster to odour stimulation than PNs (Krofczik et al. 2009) and would be candidate neurons to modulate the later responses of PNs. We hypothesise accordingly that the early responding neurons are LNs, the late responding PNs.

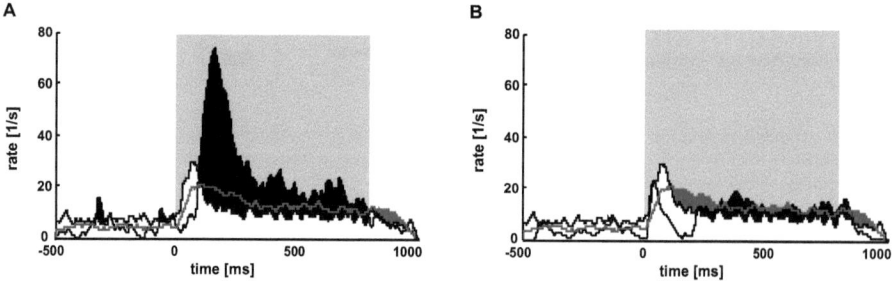

Figure 2.6: Early responding AL neurons may impact late responses in AL neurons. Group averaged rate functions of early responding 'elemental neurons' (white, n = 9), late responding 'elemental neurons' (black, n = 3) and 'configural neurons' (grey, n = 6). Light grey background indicates stimulus delivery time. The narrow late group comprised responses of excitation (A) and inhibition (B) which are treated separately. **(A)** Rate peaks of early neurons precede positive rate peaks of excited late AL neurons. **(B)** Rate peaks of early neurons also precede negative rate peaks of inhibited late neurons. Note the initial small activity boost which precedes the inhibited period.

2.3.5 Hetero LNs are involved in configural as well as elemental processing.

A cells function is determined by its physiology and morphology. We therefore analysed the morphology of a subgroup of our neurons (n=4). One of them was a PN, the other three were LNs. All four morphologies confirmed previous classification in LN/PN based on their response latencies.
Two of the LN stainings were of sufficient quality to be reconstructed. Both neurons were hetero LNs and responded to 2-heptanone and its binary mixtures. One was a narrow tuned elemental neuron and the other a configural neuron. We asked whether their glomerular innervation pattern could explain their different response profiles. We identified the innervated glomeruli and compared these

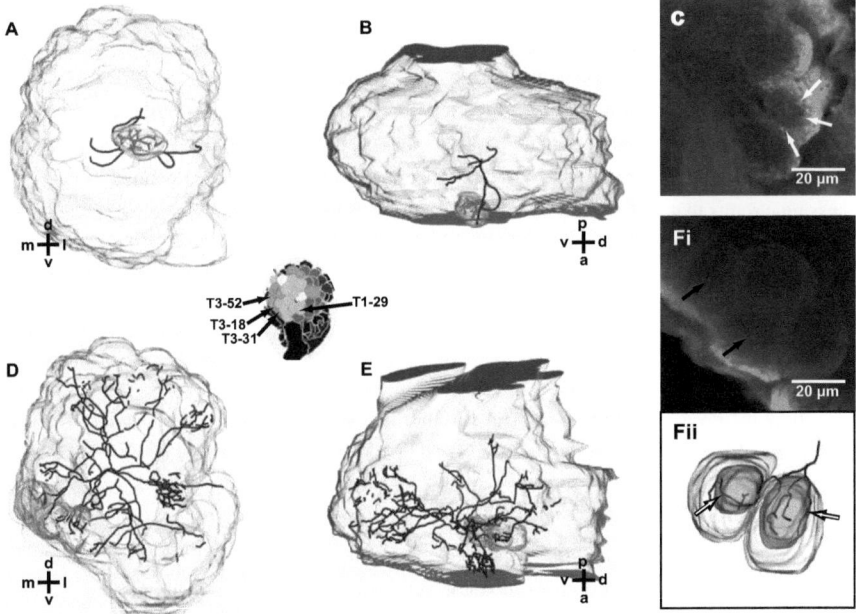

Figure 2.7: **Glomerular innervation patterns of hetero LNs responding to 2-Heptanone.** The schematic AL in the middle of the panel illustrates in grey scale the involvement of single glomeruli in the response to 2-Heptanone (white, highly active, black no activity or no data), as determined by calcium imaging with bath applied dye (cp. http//:neuro/uni-konstanz.de the physiological atlas of the honey bee.) Arrows indicate glomeruli which are innervated by the neurons presented. **(A)** Frontal view of a reconstruction from a hetero LN exhibiting elemental coding of 2-Heptanone. **(B)** Parasagittal view of the same hetero LN as in (A). **(C)** Confocal image illustrating the dense innervation of the glomerulus. Overlapping innervation area is indicated by white arrows. **(D)** Frontal view of a reconstruction from a hetero LN which coded configurally for 2-Heptanone. **(E)** parasagittal view of the same hetero LN as in (D). **(Fi)** Confocal image illustrating the sparse innervation of glomeruli in the core region. Sparse arbors seem not to overlap with ORNs (black arrows). **(Fii)** reconstruction of glomerular cap and core as well as sparse arbors from Fi. Note that neurites distribute just between cap and core (black-white arrows)

with the AL's spatial activity pattern evoked by 2-Heptanone as published in the physiological atlas of the honey bee (http://neuro.uni-konstanz.de). It turned out that the densely innervated glomerulus of the elemental neuron was one of the 2-Heptanone responsive glomeruli (T1-29, Fig.2.7 A,B). The neurites branched within the core of the glomerulus and reached out into an intermediate layer between cap and core. Counterstaining of ORNs showed that LN branches and ORN axons overlapped suggesting a direct connection (Fig.2.7 C, white arrows). The main glomerulus of the configural neuron, however, innervated a glomerulus that is not responsive to 2-Heptanone (T1-19) but the same neuron innervated three glomeruli sparsly that are weakly responsive to 2-Heptanone (T3-18, T3-31, T3-52, Fig.2.7 D,E). Sparse arborisations was not apparent in the glomerular cap (Fig.2.7 Fi, black arrows). However, a careful reconstruction of the neuron and the glomerular cap based on counterstained ORNs showed that the sparsely arborising neurites in fact distributed just between cap and core (Fig.2.7 Fii, black-white arrows). Hence, sparse arborisations could equally allow for direct, monosynaptic ORN input from the cap and for poly-synaptic input through LNs and PNs from the core.

While the response latency of the elemental neuron to the dominant odour and its mixtures was similar (36 ± 2 vs. 38 ± 4 ms), the configural neuron clearly responded faster to the single compound (18 ± 1 vs. 49 ± 11 ms). This change in latency indicates the occurrence of both, mono- and poly-synaptic input to the configural processing LN. We propose that this neuron was embedded in two different processing circuits that were differentially activated depending on the sensory stimulus delivered.

2.4 Discussion

In this study we investigated physiological and morphological properties of honey bee AL neurons, LNs in particular. For this purpose, we conducted parallel intracellular and EAG recordings under stimulation of mono-molecular odorants and their temporally perfect and imperfect binary mixtures. We made use of the phenomenon of concentration-dependent latency shifts, which are generated in the antenna, to separate stimulus specific response latencies in AL neurons. This novel approach allowed us to classify odour-mixture responses of single neurons, identifying whether a neuron responded to a compound of the mixture or to the mixture as a new odour. The EAG signal provided an internal reference time point that allowed to deduce latencies of AL neurons unambiguously. We found that half of the neurons responded to one of the compounds rather than the mixture and we termed these 'elemental neurons'. These neurons clustered into narrow tuned LNs, broadly tuned LNs, and narrow tuned PNs. The other half of the recorded AL neurons exhibited largely individual responses, both, in terms of tuning and latency. However, they had in common that responses to stimulations with mixtures did not reflect feature information. We termed these neurons 'configural neurons'. Configural neurons expressed a tendency to respond on the individual scale faster to temporally perfect stimuli, implicating that they were engaged in different processing circuits, depending on stimulus context. This assumption is further strengthened by morphological evidence which implicates that hetero LNs can facilitate elemental as well as configural coding.

2.4.1 Broad and narrow tuned LNs - functional subgroups?

In this study, we classified neurons as broadly or narrowly tuned depending on whether they responded to one or both of the single components. While useful in the context of this study it is clear that a "narrow" neuron may turn out to be a broadly tuned neuron for another set of tested stimuli.
Nevertheless our observation is not without interest since it shows LNs to be odour specific to a certain degree. Odour specificity becomes apparent by the presence or absence of a response in narrowly tuned LNs but also by subtle odour specificity of broadly tuned neurons as visible from their temporal response properties. Subtle odour specificity has also been shown in Drosophila (Wilson and Laurent 2005). In previous investigations, honey bee LNs have been reported to respond to odours rather unspecifically (Abel 1997; Krofczik et al. 2009; Sun et al. 1993). Odour concentrations used in those studies were probably higher than used in our experiments, even if multiple conditions influencing the actual amount of odour arriving at the antenna are taken into account. It seems reasonable to assume that odour specificity of LN responses increases with decreasing concentration of the stimulus. This assumption is in line with results from calcium imaging that show that decreasing stimulus concentration leads to fewer active glomeruli (Carlsson and Hansson 2003; Friedrich and Korsching 1997; Sachse and Galizia 2003).

2.4.2 Puzzles of suppression and excitation - a complex AL blueprint?

Electro-physiological recordings have the advantage of high temporal resolution. We chose this method, hoping to infer a given neuron's position in the AL circuitry from its mean response latency. Our results support earlier findings according to which LNs respond prior to PNs and influence PN output decisively (Christensen et al. 1993; Krofczik et al. 2009; MacLeod and Laurent 1996; Wilson and Laurent 2005; Wilson et al. 2004).
Single cell recordings in moth have revealed clustering of LNs into groups of different latencies suggesting successive LN to LN activation (Christensen et al. 1993). In agreement with previous work (Flanagan and Mercer 1989; Sun et al. 1993), we could not show any such obvious clustering for LNs of the honey bee. Still, our finding of one hetero LN in particular and the general tendency of configural neurons to respond with longer latencies to mixtures as compared to a single compound is best explained by LN-LN interaction which occurs in a stimulus-context dependent manner. Our results further reveal at least two different types of LN-PN interactions. Delayed information transmission resulting in a phasic-tonic PN excitation and late inhibition of PNs. What mechanisms could underlie these LN-PN interactions? Delayed information transmission could either be realized through excitatory LNs, or a disinhibitory ORN-LN-LN-PN pathway. Both of these mechanisms have been shown to exist in insects (Christensen et al. 1993; Huang et al. 2010; Shang et al. 2007; Yaksi and Wilson 2010), but direct evidence from the honey bee is still lacking. Late inhibition of PNs could be realized by an inhibitory ORN-LN-PN pathway or by recurrent inhibition between PN and LN, as in the mammalian olfactory bulb (for review see Urban and Arevian 2009). In both cases the PN would be excited by an ORN and only subsequently inhibited by an LN that was either activated by

the same ORN or by the PN itself. This would explain the small activity boost prior to the inhibition (Fig. 2.6 B). Clearly, the honey bee AL follows a complex blueprint that combines more than one functional network to resolve its task of olfactory coding. More anatomical and electro-physiological evidence is needed to understand these networks.

2.4.3 Hetero LNs - multi-function neurons?

Local neurons of asymmetric shape are a common feature of the insect AL (Chou et al. 2010; Christensen et al. 1993; Flanagan and Mercer 1989; Seki and Kanzaki 2008; Stocker et al. 1990; Sun et al. 1993). Nevertheless, hetero LNs with their densely innervated main glomerulus are common only amongst a few species of hymenoptera (Dacks et al. 2010) and best known from the honey bee. What could make this oddity be of particular use to the honey bee?

Hetero LNs have been speculated to be functionally polar neurons. Polarity could be realized in one of two possible ways: Hetero LNs could receive input in their densely innervated glomerulus (focused input), or hetero LNs could give output to their densely innervated glomerulus (focused output). Evidence for focused input is derived from imaging experiments. Most, but not all, hetero LNs respond in accordance with the odour response profile of the PNs in their prominent glomerulus. This layout of a focused input would make hetero LNs useful in shaping intra-glomerular inhibition and elemental mixture processing (Galizia and Kimmerle 2004). The neuron would gather information directly from the ORNs innervating the dense glomerulus und restrain the activity of other glomeruli by targeting PNs through inhibitory synapses in the glomerulus core. Amongst the neurons we recorded and classified as LNs featuring elemental processing, we had at least one example supporting the focused input hetero LN layout (Fig.2.7 A-C). Support for focused output (Lei and Vickers 2008) comes from LN development showing pruning in a dendrite-like fashion in sparse but not dense arbours (Devaud and Masson 1999). This layout would make hetero LNs useful tools in configural mixture processing. The neuron could collect information, not from ORNs, but from PNs and LNs of many active glomeruli and potentially inhibit PNs or ORNs of the densely innervated glomerulus. This could explain the few mismatches of LN response and glomerulus profile, which were shown by calcium imaging (Galizia and Kimmerle 2004). Data presented here also deliver an example of a configural hetero LN supporting the focused output view (Fig.2.7 A-C).

Having data at hand to support both of the mutually exclusive possibilities, the question arises whether the concept of polarity really applies to the hetero LNs. Possibly, hetero LNs resemble the vertebrate olfactory granule cell that perform complex signal processing based on dendrodendritic interaction (Shepherd et al. 2007). For these interactions active membrane properties and reciprocal synapses are prerequisites. In the cockroach, input and output synapses are present on the same neurite in the same glomerular substructure (Distler et al. 1998; Malun 1991a). Evidence for reciprocal synapses also exists in the honey bee AL (Gascuel and Masson 1991). Furthermore, multiple spike heights have repeatedly been described in LN recordings from honey bees and moths indicating the possible existence of multiple spike initiation zones (Christensen et al. 2001, 1993; Flanagan and Mercer 1989;

Galizia and Kimmerle 2004; Krofczik et al. 2009; Sun et al. 1993). Taken together, hetero LNs might combine focal input and focal output.

Hetero LNs as multi-functional neurons could explain why these neurons are so abundant in honey bee but not in other insects. Honey bees rely to a greater degree on a flexible odour encoding system than e.g. moth or Drosophila. This is, on the one hand, because they have to learn the odours of many flowering plants and localize these. On the other hand, and maybe more importantly, because their complex social interaction relies on pheromone communication, but apparently lacks labelled lines, as we know them from other insects for purposes of intra-species comunication(Sandoz et al. 2007). As a consequence, they have to have mechanisms to tell apart compositions in which an odorant serves as a pheromone from those where it is the mere constituent of a floral fragrance. A multi function hetero LN could serve as a means to evaluate elemental and contextual aspects of odours.

CHAPTER 3

Local Interneuron Morphology

Contents

- 3.1 Introduction 23
- 3.2 Materials and Methods 25
  - 3.2.1 Animal preparation 25
  - 3.2.2 Stainings and morphological preparation 25
  - 3.2.3 Confocal imaging and data processing 27
- 3.3 Results 28
  - 3.3.1 Inter-glomerular innervation patterns 28
  - 3.3.2 Intra-glomerular arborisations 29
  - 3.3.3 Branching shape of the main neurite 30
  - 3.3.4 Morphological diversity of LNs 31
  - 3.3.5 Phenotype and neurite thickness 33
- 3.4 Discussion 33
  - 3.4.1 The LN attitude - PNcentric, or ORNcentric? ... 33
  - 3.4.2 Just homo or hetero? - LNs are morphologically diverse. ... 34
  - 3.4.3 Means of communication - possible assignment of neuro-transmitters and -peptides to the described LN phenotypes. ... 35

3.1 Introduction

Neurons within the central nervous system can be divided into two principal classes. Projection neurons (PNs) connect different brain areas whilst local neurons (LNs) are spatially restricted to one brain area. Together these neurons assemble multiple parallel circuits in which information about the environment is transformed and processed such that the organism can exhibit an appropriate behaviour. Obviously, in order to understand information processing it is important to know about the properties of single cells involved as well as the wiring between cells. In this light, characterisation of LNs on a morphological basis is a prerequisite to understand nervous system information processing.

In the honey bee antennal lobe (AL), the first olfactory neuropile, $\sim$2400 LNs outnumber $\sim$800 PNs about threefold (Bierfeld 2009). LNs establish connections within and between the 160 spherical subunits of the AL, the glomeruli. Each glomerulus receives sensory input by $\sim$400 olfactory receptor neurons (ORNs) but about double the amount of LNs (Galizia 2008). These numerical relations alone suggests that LNs perform multiple important tasks in the AL circuitry.

In line with this suggestion, insect LNs are a diverse class of cells differing in morphology (Chou et al. 2010; Christensen et al. 1993; Dacks et al. 2010; Flanagan and Mercer 1989; Fonta et al. 1993; Seki and Kanzaki 2008; Seki et al. 2010; Stocker et al. 1990), expression of neurotransmitters and peptides (Barbara et al. 2005; Berg et al. 2007; Carlsson et al. 2010; Dacks et al. 2010; Homberg 2002; Kreissl et al. 2010; Nässel 1999; Nässel and Homberg 2006; Schäfer and Bicker 1986), as well as their electrical properties (Chou et al. 2010; Husch et al. 2009b). In terms of neuritic morphology, honey bee LNs get divided into two types: homogeneous (homo) LNs and heterogeneous (hetero) LNs (Flanagan and Mercer 1989; Fonta et al. 1993). Homo LNs have a uniform appearance and innervate many glomeruli sparsely. Hetero LNs are given a non-uniform appearance by densely innervating a single glomerulus and many others sparsely. The variability of phenotypes within these two groups remains large, suggesting that each of the two groups might be an assembly of several functional subgroups (Chou et al. 2010; Dacks et al. 2010; Seki and Kanzaki 2008; Seki et al. 2010).

A means to subdivide LNs on a functional basis is to identify the targets of their neurites (Markram et al. 2004). In case of the AL, neurite targets can be reasonably defined on a global level of inter-glomerular innervation within afferent and efferent AL sub-fields and on a local level of intra-glomerular arborisation between input and output layer. Both levels are established by the distribution of ORN and PN neurites:

Within each glomerulus, ORN and PN neurites are sorted in a layered fashion. ORN output synapses distribute throughout the superficial "cap" of a glomerulus establishing the input layer. PNs establish the output layer. Honey bee PNs typically invade either the central "core" of several glomeruli (multiglomerular PNs), or cap and core of only a single glomerulus (uniglomerular PNs) (Fonta et al. 1993). Honey bee ORNs enter the AL through four sensory tracts (T1-4) which segregate the AL in four afferent sub-fields by innervating discrete sets of glomeruli (Suzuki 1975). Each glomerulus is thought to receive input from only one type of ORN expressing one type of receptor (Robertson and Wanner 2006; Sachse and Galizia 2006; Vosshall et al. 2000) such that a glomerulus has a unique response spectrum (Carlsson et al. 2002; Galizia and Menzel 2000). Accordingly, the four discrete afferent sub-fields may have functional significance (for review see Galizia 2008). Uniglomerular PNs leave the AL via two antenno-cerebral tracts (ACT), medial (m-ACT) and one lateral (l-ACT). Again, each glomerulus is innervated by PNs projecting through the same ACT such that the AL is similarly segregated into two efferent sub-fields (Abel et al. 2001). These efferent sub-fields coincide fairly well each with two of the afferent sub-fields, dividing the AL into two hemilobes and establishing a dual olfactory pathway (Kirschner et al. 2006). Accumulating evidence for different response profiles of m-ACT and l-ACT PNs (Krofczik et al. 2009; Müller et al. 2002) suggests a functional significance also for the dual olfactory pathway (Galizia and Rössler 2010).

It becomes clear that we have quite a good understanding about the ORN - PN connectivity in the honey bee AL. Still, little is known about the position LNs take within this blueprint. Do innervation patterns of LNs show a preference to orient towards either afferent or efferent sub-fields on the global

AL-scale? And on a local scale, do dense and sparse arbours overlap differently with cap and core of the glomerulus? How do local arborisations meaningfully complement global innervation patterns? And is there a system underlying the variability of LN morphologies?

To approach these questions, I reconstructed morphologies of single neurons from stainings of different LNs. By transforming these neurons to fit into a common reference frame it was possible to evaluate which afferent and efferent sub-fields received innervations. To analyse intra-glomerular arborisation I reconstructed cap, and where possible, core of single glomeruli as well as sparse and dense arborisations of the LNs innervating them. Based on these and additional morphological descriptors I attempt to group LNs into meaningful phenotypes.

3.2 Materials and Methods

3.2.1 Animal preparation

Worker honey bees (*Apis mellifera*) were caught at the entrance of the hive or at a feeder, immobilized by cooling, and mounted in custom made plexiglas holders. The bees were allowed to acclimate to the new environment for 1-6 hours before the experiment started.

Antennae were immobilized with Eicosane (melting point 37 C°; Sigma-Aldrich Chemie GmbH, Germany), whilst head and mandibles were immobilized with Deiberit 502 (melting point 60 C°; Boehme-Schoeps, Germany). To reduce brain movements, the oesophagus was detached from its muscles. The head capsule was opened between the median ocellus and the base of the antennae. Glands and tracheal sheaths were removed carefully. The exposed brain was kept moist during recordings by dribbling saline onto it, if necessary (in mmol $_{-1}$: 130 NaCl, 6 KCl, 4 $MgCl_2$, 5 $CaCl_2$, 10 Hepes, 25 D-Glucose, 160 sucrose, pH 6.7, 500 mosml).

3.2.2 Stainings and morphological preparation

I wanted focus attention on the functional morphology of honey bee LNs. For this purpose single neuron morphologies were visualised by intracellular dye application. Counter stainings of the AL neuropil allowed for access to information about the inter-glomerular innervation patterns. Backfills of ORNs were performed in some preparations. Backfills gave the possibility to analyse intra-glomerular innervation patterns. The procedures are detailed below.

Intracellular staining of AL neurons

To stain single neurons fluorescent dye or morphological marker were iontophoretically injected using sharp glass electrodes. Electrodes were pulled from borosilicate capillaries (GC150F-10, Clark electronic instruments, UK) by means of a horizontal Puller (P-97, Sutter Instruments Co.,USA) and tip-filled with injection solution. A variety of dyes and markers were used for injections, these were 2.5% Alexa-Fluor 546 Biocytin in 0.2 M KCl, 4% Alexa 488 Hydrazid in 0.2 M KCl, 4% Micro Ruby in 0.2 M K-Acetate or 3% Lucifer Yellow in 0.1% LiCl (Sigma-Aldrich) and the morphological marker 2% Neurobiotin (vector laboratories) in 0.1 M K-Acetate. Solutions were iontophoretically expelled whith the polarity of the current pulses (0.2 s width, 2 Hz, 1-4 nA) chosen according to the dye's charge. A successful intracellular filling could be achieved more often with higher currents and if the cell recovered to its initial spontaneous activity after iontophoresis.

AL-neuropil counterstaining

In order to identify the individual glomerular innervation pattern of each neuron, I performed counterstaining of the AL-neuropile. To visualize glomeruli, I used one of the following techniques. I labelled sensory tracts or performed antibody staining against GABA.
Sensory tracts of the penetrated AL were labelled with Neurobiotin (Vector Laboratories Inc., USA). For this purpose, the cuticle previously removed from the head capsule was now replaced and closed carefully again using Eicosan. The ipsilateral antenna was brought in an upright position and surrounded by a made basin from Vaseline that was filled with 2% Neurobiotin in distilled water. The antenna was cut at the scapus and the Neurobiotin was given 2-3 hours to be taken up by the cells.
GABA staining was performed only after a successful intracellular staining. Poly-clonal rabbit anti-GABA antiserum, which is in established use in the honey bee (Kreissl et al. 2010), was kindly provided by Dr. Heiner Dircksen (University of Stockholm, Sweden). Details of the procedure are given below.

Morphological preparation

Bees were perfused with 4% para-formaldehyde by injection into the thorax. Brains were removed and fixated in 4% para-formaldehyde in phosphate buffered solution (PBS) for 3 hours or at 4° over night. Subsequently, preparations were washed for at least 10, 30, and 45 minutes in PBS. To visualise Neurobiotin in those cases where sensory tracts were labelled, brains were incubated depending on emission wavelength of the single-cell marker, either in AMCA-avidin (Vector Laboratories Inc., USA), or in Streptavidin-Cy3 for at least 5 hours and washed again 3 times for at least 15, 30 and 45 minutes, respectively. The tissue was dehydrated in an ascending ethanol series, degreased for 20 min in Xylol and finally embedded into DPX mounting medium (Fluka, Sigma-Aldrich Chemie GmbH).
In cases where a single cell was nicely stained but the sensory tracts were not labelled I performed a subsequent immunostaining against GABA. For this purpose, brains were eluted from the mounting medium by bathing them in Xylol for up to five days under gentle shaking. The Xylol was changed repeatedly during this time. Subsequently, the tissue was rehydrated in an descending ethanol series and made permeable to the antibody by washing three times in PBS with the addition of 0.2% Triton X. Permeated brains were blocked over night with 0.2% BSA in PBS with 0.2% Triton X and NaN_3. GABA antiserum was diluted 1:30000 in PBS- 0.2% Triton X, with 0.2% BSA and 0.02% NaN_3 and applied for 6–8 days at room temperature. After incubation with the primary antiserum brains were washed at least 5 times for at least 2 hours each. Secondary goat antirabbit antibodies conjugated to

Alexa 633 (Invitrogen, Carlsbad, CA) were diluted 1:200 and applied for 5 days.

3.2.3 Confocal imaging and data processing

To visualise staining results stacks of images were taken by means of a Zeiss LSM 510 Meta confocal microscope (Carl Zeiss AG, Germany) equipped with a 20x/1.0 water immersion objective. Laser lines 488 (Alexa 488, Lucifer Yellow), 543 (Micro-Ruby, Cy3, Alexa 546), 633 (Alexa 633) and 2-photon excitation at 750 nm (DAPI, AMCA) were used in combination with appropriate filter settings. To view and process confocal image stacks Zeiss LSM Image Browser or Amira 5.2.1 software (Visage Imaging GmbH, Germany) was used. Neuron morphologies were reconstructed and a set of marker Glomeruli was identified and segmented. With the help of the Amira implemented method of landmark based warping, reconstructed neurons were fitted into a 3D model of the honey bee's morphological atlas (Galizia et al. 1999, http://neuro.uni-konstanz.de/honeybeeALatlas).

Neuron reconstruction

Neurons were reconstructed by manual tracing using the standard filament editor. Neurite diameters were estimated by employing the Amira implemented algorithm (ScanConvertNeuronTree), which operates based on level of brightness. In some preparations, where fluorescence of the surface was very high, this led to obviously incorrect rendering of the neuron. I had to accept this circumstance since the software does not allow warping of any other data format. Warping however was necessary to combine the single neurons into a common standard frame (c.f.: 3.2.3).

Glomerulus segmentation

Location and extend of single glomeruli were detected by interactive segmentation using the standard Amira segmentation editor. Glomerulus identity was determined by visual inspection and comparison with the morphological atlas of the honeybee (http:\\neuro.uni-konstanz.de/honeybeeALatlas). Nomenclature as established by Galizia et al. (1999) was used. Identification of glomeruli allowed for landmark based fitting of neurons from different preparations into one standard brain. In cases where labelling of GABAerg neurons was used as a counter stain, neuron reconstruction and segmentation of glomeruli were not based on parallel channels from the same scan but on two independent scans of the same preparation. In these cases reconstructed neurons and segmented glomeruli were aligned by performing a landmark based rigid warping between the two scans of the same specimen.

Landmark based warping

The glomeruli segmented and identified from each individual specimen were used to fit the reconstructed neurons into a common standard frame. Two sets of landmarks were created, one giving the location of a defined spot in the specimen, the second giving the location of the same spot in the standard 3D model. A landmark was always placed at the centre of a glomerulus. Based on these landmarks, the reconstructed neuron was transformed by means of a non-rigid warping algorithm (Bookstein) to fit into the standard 3D model. The identity of all glomeruli in this common reference frame is known which allowed examination of the neurons glomerular innervation pattern in respect to the distribution of sensory as well as antenno-cerebral tracts.

Measurement of neurite diameter

Neurite diameters were measured using the scale bar tool of the Zeiss LSM Image Browser. The thickest neurite running through the glomerulus' core was defined as main neurite. Smaller neurites branching of were defined as secondary. For each cell, the diameter of main and secondary neurite were measured at five different locations. To compare diameters between groups repeated measurement Anova was used.

3.3 Results

3.3.1 Inter-glomerular innervation patterns

By its afferent (ORN axons) and efferent (PN axons) connections, the AL sub-divides in four afferent and two efferent fields. Each of the efferent fields overlaps fairly well with two of the afferent fields, separating the AL in two hemilobes and constituting a dual olfactory pathway. I wanted to know whether LNs fit in with this organisation by preferential connection or omission of glomeruli within a certain afferent or efferent field.

For this purpose I reconstructed LN morphologies from intracellular stainings and transferred these

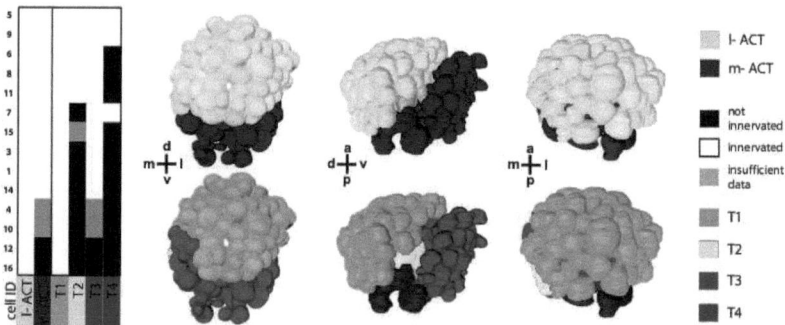

Figure 3.1: Inter-glomerular innervation patterns of individual LNs. (**A**) Grey scaled chart illustrating innervation of AL afferent fields as given by the innervation through the sensory tracts T1-4 and efferent fields as given by the innervation through l-/m-ACT PNs. White indicates innervation, black the lack of innervation, grey fields mark cases where innervation was uncertain due to incomplete filling of the neuron.
(**B**) AL efferent fields illustrated by the common reference frame. Glomeruli innervated by l-ACT PNs are shaded in light grey, those innervated by m-ACT PNs shaded in dark grey. (**C**) AL afferent fields illustrated by the common reference frame. Glomeruli of the T1 area are shaded in light grey, T2 in greyish white, T3 in dark grey, T4 in greyish black.

by means of a landmark-based fitting procedure into a common AL reference frame. Since the identity of all glomeruli within the common reference frame was known, it could easily be investigated which of the afferent and efferent fields received innervations by an individual LN (Fig. 3.1).
In this dataset, every LN invaded the T1 area (n = 14) and most invaded likewise the T3 area (n = 10). The T2 area was often (n = 8), the T4 area almost always (n = 11) omitted. It is of particular interested to know whether a notable amount of LNs would restrict their innervations to one of the two efferent fields and thereby integrate into the hemilobal structure constituting the dual olfactory pathway. I found no incidence where glomeruli of the m-ACT efferent field were innervated exclusively but some cases (n = 2) in which only glomeruli of the l-ACT efferent field were innervated. These LNs however restricted their innervations to glomeruli which belong to the T1 area.
Reviewing these results, I hypothesize that LNs establish connections within and between afferent fields, rather than between efferent fields.

3.3.2 Intra-glomerular arborisations

Each glomerulus receives dense ORN input in its cap. Accordingly, a functional connection between different glomeruli on the level of the input would require that neurites of a connecting LN would spatially overlap with the cap of each innervated glomerulus. Assuming that inter-glomerular innervation patterns of LNs predominantly establish connections between the ALs afferent fields, I next asked whether the functionality of this layout would be supported by LN intra-glomerular arborisation.

To answer this question I studied LN arborisations within both, densely and sparsely innervated glomeruli. To access intra-glomerular arborisation I reconstructed the main branches of the neurite as well as the outer boundaries of the glomerulus. In cases where a sensory backfill was used as a counter stain it was even possible to reconstruct the glomerular core.

I found two distinct patterns of arborisation in densely innervated glomeruli of hetero LNs and two different patterns in sparsely innervated glomeruli (Fig. 3.2). Dense arborisations were either fist-like wrapping around the glomerulus core (Fig. 3.2 A) or tree-like branching throughout the entire glomerular volume (Fig. 3.2 B). Sparse arborisations had often fork-like appearance, invading the intermediate layer between cap and core(Fig. 3.2 C, left glomerulus) or intermediate layer and core (Fig. 3.2 C, right glomerulus). These types of sparse arborisations occurred in homo as well as hetero LNs. Variations of fork-like inervations were non exclusive but could appear in neighbouring glomeruli that were invaded by the same neuron (Fig. 3.2 C). A second pattern of sparse intra-glomerluar arborisation could be found in one particular case of a homo LN (Fig. 3.2 D) where plenty of fine neurites spread throughout the entire glomerular volume. All observed types of intra-glomerular arborisations invaded the intermediate layer between cap and core allowing for spacial overlap between ORNs and LNs on the one hand, and LNs and PNs on the other hand.

On this morphological basis, LNs have the potential to establish functional connections between AL afferent fields and relay the gathered information onwards to PNs.

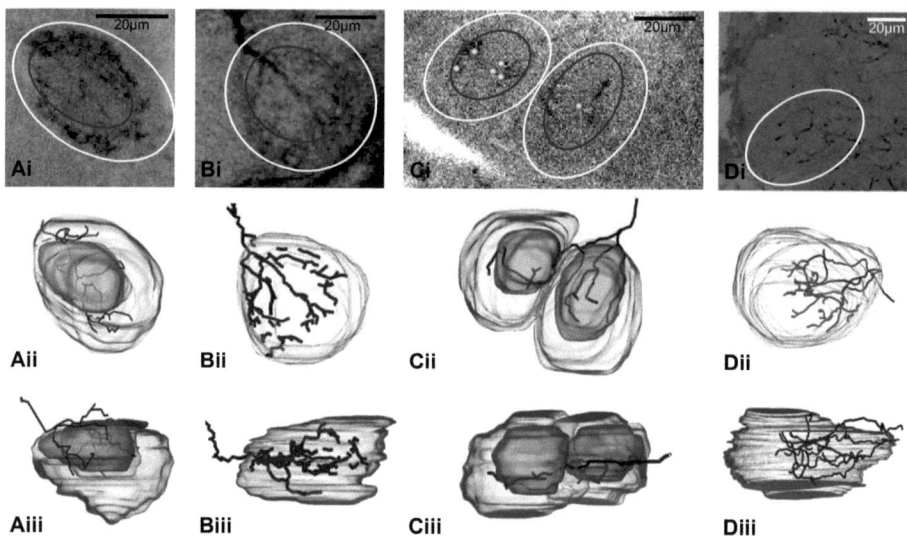

Figure 3.2: Intra-glomerular arborisations within different glomeruli. (A) Fist-like dense arborisation wrapping around the core of the glomerulus. (B) Tree-like dense arborisation spreading throughout the entire gomerular volume. (C) Fork-like arborisation invading the intermediate layer between cap and core (left glomerulus) or intermediate layer and core (right glomerulus). **D** Sparse arborisation throughout the entire glomerular volume by a homo LN.

3.3.3 Branching shape of the main neurite.

Inter-glomerular branching and intra-glomerular arborisation are two descriptors of morphology which rather directly reflect potential functions of the neuron. However morphological variability arises already from much more superficial descriptors. As such, I found the shape of LNs main neurites to be eye-catching characteristic. Main neurites in the present LN dataset could always be categorized as being either of a curled or a stellar shape (Fig. 3.3).

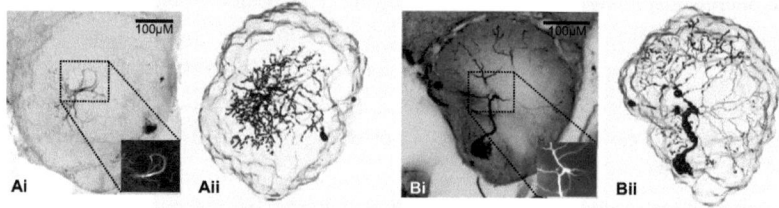

Figure 3.3: Branching shapes of main neurites in different LNs. (A) Homo LN with a curled main neurite. (B) Hetero LN with a stellar branching main neurite. Transparent red overlays in the projection view emphasise the main neurites shape.

Projection views (Fig. 3.3 Ai) and reconstructions (Fig. 3.3 Aii) of LNs with a curled branching main neurite gave often a chaotic impression. Projection views (Fig. 3.3 Bi) and reconstructions (Fig. 3.3 Bii) of LNs with a stellar branching main neurite on the opposite had a more ordered appearance. Homo as well as hetero LNs come in both shapes. However, homo LNs seemed to be more often of the curled (n = 3) than of the stellar (n = 1) type.

3.3.4 Morphological diversity of LNs

As laid out above, LN morphologies differ in several aspects and show high variability on an individual scale. I wanted to know if a categorization of predominant LN properties would allow to sort LNs according to phenotypes sharing the same characteristics.
For this purpose I used the above described characteristics:

a) innervation homogeneity (homo/ hetero LN),
b) inter-glomerular innervation pattern (innervation of afferent fields T1-4 and efferent fields l-ACT/m-ACT hemilobe, respectively),
c) intra glomerular arborisation (fist-like, tree-like),
d) location of densely innervated glomeruli (T1-4, l-ACT/m-ACT hemilobe),
e) branching shape of the main neurite (curled/stellar)

and classed every LN from of the data set accordingly.
Based on these criteria six LN phenotypes, can be differentiated in the present dataset (Fig. 3.4). Homo and hetero LNs distributed in each three different phenotypes.

Phenotype 1 (fig. 3.4 B1; n = 1) was a homo LN that innervated all four AL input fields and, accordingly, both hemilobes. Remarkably, the innervation went even below the T4 glomeruli and extended into the dorsal lobe. Intra-glomerular arborisation was of an unusual type with many small neurites invading the entire glomerular volume (c.p. Fig 3.2 D). The main neurite was of a curled shape and remarkably thick (3.2 ± 0.76 μm).

Phenotype 2 (fig. 3.4 B2; n = 1) resembled phenotype one in its homogeneous innervation of all four AL input fields. However, this neuron did not extend neurites into the dorsal lobe. Intra-glomerular arborisation was, as for most sparsely innervated glomeruli, fork-like. The main neurite branched in stellar fashion (thickness = 1.4 ± 0.25 μm).

Phenotype 3 (fig. 3.4 B3; n = 2) homo LNs omitted the T2 and T4 area. The rather fine secondary neurites (0.7 ± 0.19 μm) got even thinner when entering the glomeruli so that these neurons did not get entirely filled. Where an intra glomerular arborisation was visible, it was of a fork-like structure. The main neurite was of curled appearance.

Phenotype 4 (fig. 3.4 B4; n = 4) was most abundant in the present dataset. These neurons usually omitted the T4 area (3 of 4) and one branched exclusively within the T1 area. The fist-like, densely innervated glomerulus was always located in the T1 area. The main neurite branched in a stellar fashion.

Phenotype 5 (fig. 3.4 B5; n = 3) were hetero LNs which largely omitted the T4 as well as the T2 area. Their tree-like innervated main glomerulus was located in the T3 area. The main neurite again branched in a stellar fashion.

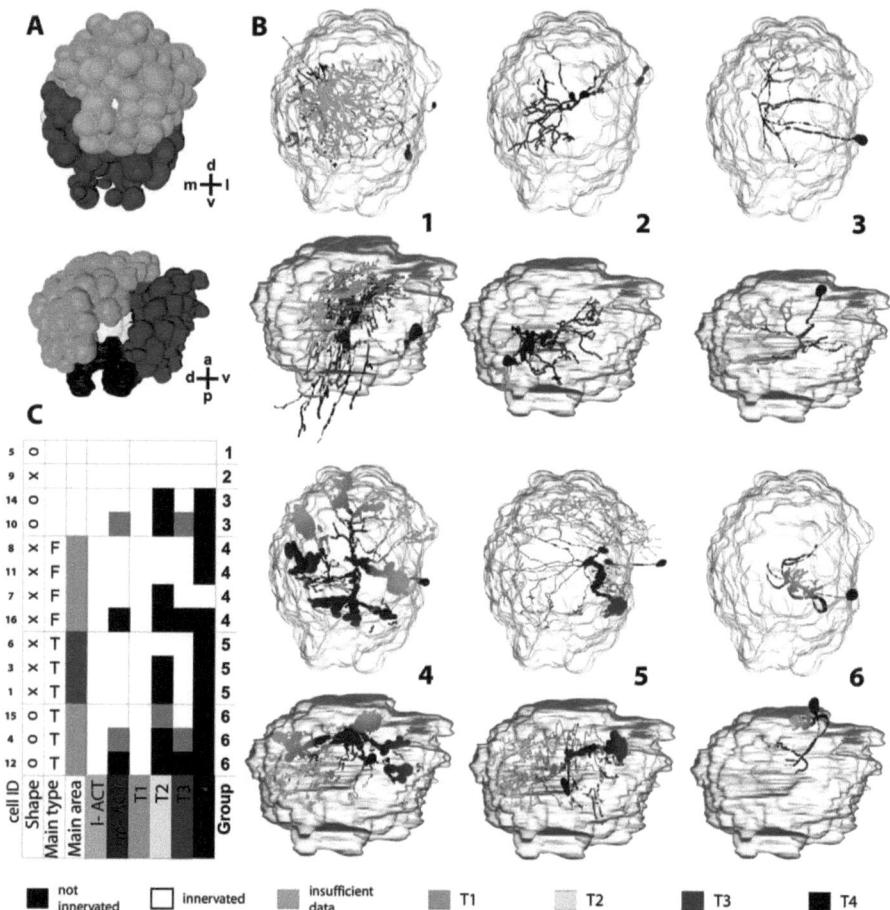

Figure 3.4: Six different LN phenotypes (A) AL afferent fields illustrated by the common reference frame. Glomeruli of the T1 area are shaded in red, T2 in green, T3 in purple, T4 in blue. **(B)** Examples of LN morphologies from each of the six phenotypes (1-6). Grey scale indicates afferent sub-field association of the innervated glomeruli. Enlarged neurites, most apparent in Phenotype 4, are artefacts arising from the requirments of the landmark based warping procedure. **(C)** Summary chart giving the phenotypic characteristic for each of the selected morphological descriptors. Grey scale code as explained in the legend, T/F = tree-like/fist-like arborisation in densely innervated glomeruli, o/x = curled/stellar shape of the main neurite.

Phenotype 6 (fig. 3.4 B6; n = 3) were hetero LNs which again omitted T4 as well as T2 innervated glomeruli and in at least one case were restricted to the area of T1 innervated glomeruli. Their tree-like innervated main glomerulus was in all cases located in the T1 area. Their main neurites branched of in a curled manner. Like for the homo LNs of phenotype 2 these neurons had rather thin secondary neurites (0.7±0.21μm) and seemed sometimes not entirely filled.

3.3.5 Phenotype and neurite thicknes

The incomplete fillings of the curled phenotype 2 and 6 led me to ask whether the neurites of curled LNs were generally thinner than those of stellar branching LNs. However this was not the case. Using stellar and curled shape as the only factor, the neurite difference in thickness was not more than a trend ($F = 0.17$). Only in a context where shape as well as phenotype were considered, neurite thickness turned out to be significantly different ($F < 0.01$).

3.4 Discussion

The honey bee AL is a highly structured neuropil, subdivided in two efferent and four afferent sub-fields established by PN and ORN axons, respectively. In the work at hand I presented an approach to place LNs into this framework and found that LN morphology is more closely related to structures defined by ORNs than by PNs. Continuing from a perspective of functional morphology, I assembled a toolbox of morphological descriptors based on which I differentiate six LN-phenotypes.

3.4.1 The LN attitude - PNcentric, or ORNcentric?

The dual olfactory pathway describes the almost perfect resemblance of the efferent AL sub-fields, with two of the afferent sub-fields each and is as such defined on the basis of PN projections. Morphological (Abel et al. 2001; Kirschner et al. 2006) and physiological (Krofczik et al. 2009; Müller et al. 2002) evidence from uniglomerular PNs gathering to l- and m-ACT suggests that the two streams perform parallel processing of different stimulus properties (Galizia and Rössler 2010). Detailed analysis of LN morphologies provides valuable information to resolve this question since it describes the building blocks of the network that shape PN tuning properties.

In line with previous descriptions of LN morphology (Abel 1997; Flanagan and Mercer 1989; Fonta et al. 1993; Kroker 2008), I found many neurons that connected the streams of the dual olfactory pathway, and some few that were restricted to one of the hemilobes. In a PN-centric interpretation, these latter neurons complement the dual olfactory pathway. However, a considerably larger amount of inter-glomerular LN innervation patterns fit meaningfully into the afferent subdivision of the AL as defined by ORN projections. In line with this notion, fork-like, sparse innervations appear to overlap with both, cap and core, such that synaptic connections between ORN and LN might exist in densely as well as sparsely innervated glomeruli. I hypothesise that ORNs are the main providers of input to at least a large population of LNs. How could LNs, which primarily relate to ORN projections be helpful in the construction of a functional, parallel processing AL network?

The two hemilobes are not mirror-reversed duplicates of the same structure but quite the contrary, since each glomerulus has a unique response spectrum. In a scenario where the dual pathway would be a perfectly segregated system, each hemilobe would have access to only that part of olfactory information which is covered by the sum of its ORN response spectra. In order to perform the sort of parallel processing that m- and l-ACT PNs are suggested to do, e.g. odour timing and odour quality (Müller et al. 2002), or configural and elmental odour-mixture information (Krofczik et al. 2009; Sun et al. 1993), a redistribution of information carried by the ORNs is necessary. LNs that branch in respect to the structure defined by the AL input could do exactly that; gather information from glomeruli of different response spectra, integrate and pass it on to PNs or other LNs.

3.4.2 Just homo or hetero? - LNs are morphologically diverse.

Honey bee LNs are usually distinguished as being either homo- or heterogeneous. While these are obvious and therefore good categories to start with, they do not describe the true variability of LN morphology. Even small sample sizes reflect large individual differences and in depth description could most likely render every single neuron in its own unique category. Without functional implication, morphological description beyond that what is most obvious is prone to be the result of subjective, rather than objective judgements. The descriptors based on which I identified the six phenotypes presented here, originate mostly in functional assumptions and their characteristics have been mentioned in previous investigations.

As one important descriptor of neuron morphology I used inter-glomerular innervation patterns. The assumption underlying the functional relevance of this descriptor, lies in the potential different information content processed by the afferent and efferent sub-fields of the AL (for reviews see Galizia 2008; Galizia and Rössler 2010). The restriction of LNs to certain parts as well as the more prominent phenomenon of cross-branching throughout the AL has been reported in every detailed description of LN morphology (Abel 1997; Flanagan and Mercer 1989; Fonta et al. 1993; Kroker 2008). In respect to the potential functional differentiation of afferent sub-fields, the invasion or bypassing of T4-glomeruli has previously been used to distinguish a sub-group of homo LNs, (D(M)Ho Flanagan and Mercer 1989). T4-ORNs are associated with the processing of not strictly odour-related information contend aspects (Abel et al. 2001; Winnington et al. 1996). Accordingly, it is quite reasonable to assume that neurons which innervate this particular area may serve cross-modal rather than pure olfactory processing and form a distinct sub-group. The particular population of homo-LNs mentioned by Flanagan and Mercer (1989) resembles the neuron I described as Phenotype 1.

Another important descriptor used by me was intra-glomerular arborisation. The functional relevance of this descriptor lies in the simple fact that cells can only establish synapses, and thereby exchange information with other cells they have spatial overlap with. While the types of dense innervation, I termed tree-like and fist-like, have previously been reported to exist in honey bee and closely related Hymenoptera (Abel 1997; Dacks et al. 2010; Flanagan and Mercer 1989; Fonta et al. 1993; Kroker 2008), my observation of sparse arbours to overlap with the ORN axons of the innervated glomerulus is in opposition to the what is commonly assumed. This seeming dichotomy might resolve itself when considering how sub-structures of a glomerulus are defined. Sparse arbours described here enter the core and branch in the intermediate layer where cap and core meet. The traditional view on glomerulus organisation arises from the existence of a concentrated innervation by ORNs in the glomerular periphery, which is called cap. The core on the other hand is solely defined by it 'not being cap'. Accordingly, it is only reasonable to assign every structure that is not most definitely innervating the cap, to the core. This dichotomy has to be called into question especially since reports of neuron morphologies with fist-like innervations, which concentrate just where cap transforms to core, came to the fore (Dacks et al. 2010; Kroker 2008). These findings give an importance to the cap-periphery which previously was neglected. Reviewing illustrations of sparse arborisations from previous works, while taking the existence of an intermediate layer into account, the innervations seemed not necessarily to be restricted to the glomerular core. Another argument in the question of sparse arbour localisation is that they have been paid little attention in comparison to dense arbours. Neurites which innervate glomeruli sparsely, are usually very thin and often not well stained. In consequence, they are more difficult to investigate. This work describes for the first time the existence of three different types of sparse innervations. Given the amount of variability LNs express and the complexity of the AL network, more types of sparse arborisations might exist, some of those perhaps truly restricted to the glomerular core.

While inter-glomerular innervation pattern and intra-glomerular arborisation reflect potential functions of the neuron, the shape of the main neurite is primarily a superficial distinction. Under the assumption that differently shaped cells serve different purposes, curled and stellar appearance of LNs might nevertheless be an interesting observation. Especially, since stellar and curled appearance correlated with type and location of a hetero-LNs dense innervation. Similar variations of the main neurite's shape have been described for LNs in the moth (Matsumoto and Hildebrand 1981).

The descriptors I suggested here may turn out to be neither complete, nor the most useful ones. Nevertheless they led to a fine scaled differentiation into six different LN phenotypes. Taken together, on an individual basis LNs express a broad variety of morphologies, but never the less may be grouped into different phenotypes. These phenotypes are distinguished by principles of organisation and hence may be indicative for functional LN subgroups.

3.4.3 Means of communication - possible assignment of neuro-transmitters and -peptides to the described LN phenotypes.

Functional grouping cannot be achieved by studying neuron morphology alone. Functional properties arise from the interplay between morphology, physiology and histochemistry. In order to define truly functional subgroups of LNs, it would be ideally to know in what combinations these properties occur. The honey bee brain displays immunoreactivity to several neuro-transmitters and -peptides (for reviews see Bicker 1999; Nässel and Homberg 2006). Visualisation of a substance's distribution results in characteristic stainings. These stainings give implications to the morphology of neurons that are equipped with the particular neuro-active substance. Based on what is known about LN histochemistry from the literature, are any of the six LN-phenotypes described here candidates to be equipped with a certain transmitter or peptide?

A small population of about 20 AL-LNs displays immunoreactivity to Alatostatin (AST; Kreissl et al. 2010). The typically large somata of the cluster termed AST6 resemble in size and position those of Phenotype 1, which is likely to be identical with the D(M)Ho group (Flanagan and Mercer 1989). AST6-Neurons, Phenotype 1 and D(M)Ho all are homo LNs with marked invasion of T4 glomeruli. While Phenotype 1 and D(M)Ho LNs extend somewhat below the AL into the dorsal lobe this was not reported for AST6 LNs. Still, the described intra-glomerular arborisations are in fair coincidence with each other such that the amount of resemblances renders Phenotype 1 a candidate for AST immunoreactivity. Since AST in the AL is always co-localised with GABA (Kreissl et al. 2010), these neurons may be expected to act inhibitory.

Histamin-like immunoreactivity is more pronounced in the AL of the honey bee than in most other insects (Nässel 1999). The transmitter may, next to GABA, act as a second potent inhibitor in the bee brain (Sachse et al. 2006). Recent investigation of histamin-like immunoreactive LNs in the bumble bee (Dacks et al. 2010) revealed these neurons as hetero LNs with fist-like, dense innervations, comparable to Phenotype 4. The pattern of histamin-like immunoreaktivity in the AL of the honey bee (Bornhauser and Meyer 1997) resembles the one described for the bumblebee in that glomeruli appear to have a ring shaped staining, most likely coinciding with the intermediate layer. Based on this resemblance, honey bee histamin-like immunoreactive neurons can be expected to be hetero LNs with fist like dense innervations.

While histamin-like immunoreactivity produces ring shaped structures, immuno stainings against the neuro-peptide Tachykinin (Bierfeld 2009) fill the entire core of glomeruli. Such a pattern could be produced either by multiple innervations from homo LNs or by innervations of at least one tree-like hetero LN in each glomerulus. Tachykinin and Histamin are co-localised in a small sub-population of honey bee LNs (Bierfeld et al. 2011). Since histaminergic LNs are likely to be hetero LNs, this finding makes it more reasonable to assume that tachykininergic LNs might likewise be hetero LNs.

Both, histamin-like and tachykininerg immunoreaktivity are found throughout the AL and not restricted to one of the hemilobes. Assuming that Phenotype 4 LNs would be histaminerg, and Phenotype 5 or 6 LNs would be tachykininerg, the coincidence of neurite shape, type of dense innervation and localisation of the densely innervated glomerulus reported here has to be an artefact of the sample size.

The question whether morphological distinct LNs are also distinct in their histochemical properties can ultimately only be resolved by parallel labelling of single neurons and immuno stainings. However, comparison between described morphologies and patterns of distribution from neuro-active substances produces educated guesses based on which targeted investigation of these questions get more effective.

CHAPTER 4
Clustering of evoked activity from antennal lobe neurons.

Contents
4.1	**Introduction**	37
4.2	**Materials and Methods**	38
	4.2.1 Data	38
	4.2.2 Data preprocessing	39
	4.2.3 Data descriptors	39
	4.2.4 Computation of descriptors	40
	4.2.5 Statistical analysis.	42
4.3	**Results**	43
	4.3.1 Clustering of AL neuron activity patterns based on spiking and sub-threshold information.	43
	4.3.2 Clustering of AL neuron activity patterns based on spiking information alone. . .	46
	4.3.3 Distribution of PNs and LNs in the different cluster trees.	48
4.4	**Discussion**	49
	4.4.1 Holistic or simplistic - how much information is necessary to distinguish meaningful clusters?	50
	4.4.2 Science or fiction - may electro-physiological characteristics be used to predict AL-neuron morphology?	51
	4.4.3 Utile or futile - why do we need established electro-physiological groups of AL neurons?	52

4.1 Introduction

Intrinsic electro-physiological properties are decisive for a cell's function (Lliñs 1988). Consequently, electro-physiological measures are, next to taxonomic measures and histochemical characteristics, established means based on which groups of neurons are classified (Connors and Gutnick 1990; Markram et al. 2004; PING et al. 2008).
Similar to the mammalian neocortex, neurons in the first olfactory neuropil of the insect, the Antennal Lobe (AL) dosplay a variety of firing patterns (Chou et al. 2010; Christensen et al. 1993; Husch et al.

2009a; Sun et al. 1993). Different from the mammalian cortex these are rarely used as a means of classification. In the honey bee (*Apis melifera*) in particular, spiking activity of single neurons has been reported to differ in the regularity of amplitude (Galizia and Kimmerle 2004), strength and regularity of spontaneous activity (Flanagan and Mercer 1989; Sun et al. 1993), nature of the preferred stimulus (Krofczik et al. 2009; Müller et al. 2002; Sun et al. 1993), response latency (Krofczik et al. 2009; Müller et al. 2002) and complexity of evoked activity patterns (Abel et al. 2001; Krofczik et al. 2009; Müller et al. 2002). Despite, or perhaps rather as a consequence of this diversity, sub-groups of AL neurons commonly relate to morphological descriptions.

Two principal classes of neurons reside in the of the honey bee AL. These are Projection Neurons (PNs) on the one hand and Local interneurons (LNs) on the other hand. While PNs send axons to higher order processing centres, LNs are of unknown polarity and restrict their neurites to the AL. According to the three Antenno Cerebral Tracts (ACT) through which they project (Mobbs 1982), PNs are most commonly further subdivided into median- (m), lateral- (l) and medio-lateral- (ml) PNs (Abel et al. 2001; Galizia 2008). Like PNs, LNs are divided in morphological sub-groups: Hetero LNs innervate a single glomerulus densely and several others sparsely, homo LNs innervate multiple glomeruli only sparsely (Fonta et al. 1993). Further subdivision based on more detailed taxonomic measures is possible (Chapter 2 Flanagan and Mercer 1989) but not commonly used.

As a result of functional investigations, morphologically distinct neurons have sometimes been attributed certain spiking patterns. In particular mPNs have been reported to exhibit evoked activity patterns different from lPNs (Krofczik et al. 2009; Müller et al. 2002). Likewise, spiking properties of LNs have repeatedly been suggested to differ from PNs, but never conclusively shown to do so (Abel et al. 2001; Flanagan and Mercer 1989; Galizia and Kimmerle 2004; Sun et al. 1993).In the present study, I explore grouping of AL neurons based on electro-physiological properties rather than morphology. For this purpose, I define quantitative descriptors of spiking and sub-threshold activity. I then collect descriptive values for a heterogeneous set of AL neurons. The multi-dimensional dataset was reduced and structured by performing Principal Component Analysis (PCA) and subsequent hierarchical clustering. Obtained clusters offer a tool to distinguish the most striking characteristics in which activity patterns of AL neurons differ. By including morphological information where possible I ask whether any of the distinguished electro-physiological activity patterns correlate with a common morphological group.

Establishing a classification of honey bee AL neurons with an electro-physiological focus will aid the investigation of AL neuron physiology.

4.2 Materials and Methods

4.2.1 Data

Analysis of odour evoked activity patterns was performed on intracellular recordings of 67 AL neurons. The data pool comprised recordings from two different laboratories, conducted by different experimenters, during different periods. Cells were recorded under various experimental paradigms using different primary odorants and mixtures. The average number of tested odours was 7 (min = 1; max = 30), the average number of responded trials was 9, (min = 3; max = 40). An odour that evoked a response in a cell once, did so in every repeated trial. Stimulus duration was either 800 ms or 2000 ms. For details of data acquisition refer to Chapter 1 or Krofczik et al. (2009). Based on intra-cellular single cell stainings some cells were identified as m-PNs (n = 12), l-PNs (n = 5), ml-PNs (n = 1) or LNs (n = 7). Further cells were classed as putative m-PNs (n = 10), l-PNs (n = 2), or LNs (n = 3) by their recording position (Krofczik et al. 2009).

4.2.2 Data preprocessing

Potent stimuli, i.e. stimuli that evoked responses, were identified for each individual cell by a human observer. Responses to potent stimuli were differentiated in excited, inhibited and single spike to allow for a response-adjusted estimation of latency (c.f.: 4.2.4). Trials were cut to 500ms pre-stimulus onset and 500ms post stimulus offset irrespective of stimulus duration. Spike-times were detected using Spike2 (Cambridge Electronic Design, UK) or custom written routines in R (http://www.R-project.org) based on the open source packages SpikeOMatic (Pouzat et al. 2004) and STAR (Pouzat and Chaffiol 2009). In order to include sub-threshold characteristics it was necessary to remove the spikes from the signal. This was done by detecting local minima flanking a spike, using a flexible window routine and replacing the signal between each pair of minima with the local mean potential, as given by the 50 ms preceding the spike. Spike-less signals were baseline corrected and normalised to values between zero and one for better comparability between recordings.

4.2.3 Data descriptors

Nine features describing different properties of neural activity were chosen to separate stereotypes of evoked responses from each other. From these features eleven descriptive values were computed for each cell as detailed below.

Mean spontaneous firing rate ('Spontaneous Rate'). The mean rate of a cell gives an estimate of the number of spikes per second a neuron produces in the absence of a driving stimulus. A low mean rate describes low spontaneous activity.

Deflection from the mean rate ('Rate Increase'/ 'Rate Decrease'). Deflections from the mean rate immediately following stimulus application are the most common definition of evoked activity. Rate Increase gives a measure for excitation, Rate Decrease for inhibition (Shinomoto 2010).

Coefficient of variance ('CV'). The CV gives a measure for the spike-time irregularity of a neuron. Practically a low CV denotes a regular spiking cell (Nawrot 2010).

Spike amplitude regularity ('Regularity'). Single cell recordings from AL neurons often exhibit spikes of different amplitude on the one hand, or extremely regular spike hight on the other hand.

Regularity of spike amplitude has sometimes been used as a discriminator between LNs and PNs (Galizia and Kimmerle 2004)

Latency A cells latency allows to approximate where in a circuit the cell is involved. Cells which perform second order processing typically have longer latencies than those which perform first order processing steps.

Variability of Latency ('Latency Variance'). The reliability of a cell's latency gives an approximation of how stereotyped the employed processing circuit is. Variability of cell latency therefore is a descriptor for the steadiness of the underlying circuitry.

Mean spontaneous signal power ('Baseline Power'). Next to spiking activity a neuron is characterised by its membrane properties. The power of a signal pre-stimulus onset, from which spikes are removed, describes how much the membrane potential fluctuates in the absence of a driving stimulus.

Stimulus related signal power ('Stimulus Power'). By subtracting the spontaneous power from the mean power of the signal during stimulation, an estimate of the stimulus related power is obtained.

Area values from spike-less signal ('Depolarisation'/ 'Hyper-polarisation'). In some instances evoked activity is accompanied by a stimulus-correlated deflection in the membrane potential. The size of the area below or above this deflection describes the strength of a de- or hyper-polarisation of the membrane, respectively.

4.2.4 Computation of descriptors

Spontaneous and stimulus correlated firing rate (Spontaneous Rate, Rate Increase, Rate Decrease).

Firing rate functions were estimated based on pooled, trial-aligned spike-trains. These were derived using a method adapted from Meier et al. (2008). In brief: First, the derivative of each single trial spike-train of a given cell under stimulation of one odour was estimated by convolving with an asymmetric Savitzky-Golay filter (Savitzky and Golay 1964) (polynomal order 2, 300 ms width, Welch windowed). Second, all single trial-derivatives were optimally aligned, finding the greatest possible pair-wise cross correlation (Nawrot et al. 2003). Third, the single-trial spike-trains were temporally aligned by shifting each by its individual delay as given by the cross-correlation. Fourth, the aligned spike-trains were merged into one train, representing the cells within-odour activity. Fifth, the alignment was repeated between the merged spike-trains of different odours. Sixth, the within-odour merged and across-odour aligned spike-trains were merged again into one single spike-train for each cell, representing its summed activity across odours.

To estimate the trial-averaged rate function the summed across-odour spike-train was convolved with an asymmetric alpha kernel ($k(t) = t^{(-t/\tau)}$) (Nawrot et al. 1999; Parzen 1962). Optimal kernel width (defined as the standard deviation of the normalised kernel function) was estimated on the basis of the empirical data, by application of a heuristic method detailed in (Nawrot et al. 1999). Mean spontaneous rate was obtained from 500 ms pre-stimulus onset. Deflections from the spontaneous rate were defined as the absolute minimum, respectively maximum within the post-stimulus intervals (stimulus-duration plus 500 ms post offset) of all trials from one cell, normalised by subtraction of the mean rate.

Coevicient of variance (CV).

The Coefficient of Variance (CV), is a classical measure of spike-time irregularity defined as the dispersion of the inter spike intervals (ISIs) (Nawrot 2010). It traditionally assumes a constant rate over time and that the variation of ISIs is of stochastic nature. These assumptions are clearly violated in stimulus-modulated activity. In order to allow for a reasonable estimate of CV in these types of data nevertheless, the CV^2 (Holt et al. 1996) is accepted as a useful and efficient local measure (Ponce-Alvarez et al. 2010). The CV^2 is the ratio m between two consecutive ISIs in a spike-train, given by $2*((|x-1|)/(x+1))$, averaged over the interval of interest. Here CV^2 was first calculated for each single trial and than averaged over all trials.

Spike amplitude regularity (Regularity).

To quantify regularity the variance of differences in normalised peak amplitude ($\hat{U}$) between succeeding spikes, given by the variance (σ) of the difference between spike peaks ($r = \sigma(\delta(\hat{U}/\bar{U}))$), was used. Normalizing amplitue rendered the regularity index dimensionless. By calculating the variance of the differences rather than amplitudes themselves, steady amplitude de- or increase originating from changes in electrical contact between cell and electrode did not affect the regularity index.

Temporal measures of response onset (Latency, Latency Variance).

Absolute latency, that is the mean latency across trials, and relative latencies, that is trial-to-trial differences in latency, were calculated with one of three methods (1-3). The method was chosen based on the respective firing pattern. 1) Latencies of cells which responded to stimulation with an increased firing-rate were estimated based on the derivative of the trial-aligned firing rate as described elsewhere (Meier et al. 2008). The procedure of trial alignment was conducted as described above. The individual shifts for each trial correspond to their relative latencies. Their standard deviation σ gives a measure for the across-trial latency variability. By convolution of the summed across-odour spike-train with the same asymmetric Savitsky-Golay filter that was used for the alignment procedure, an estimate about the derivative of the cell's absolute firing rate was obtained. The cell specific absolute latency was defined as that point in time where the slope of the firing rate was steepest, i. e. the derivative's maximum. 2) Latencies of cells which responded to stimulation with a decrease in firing rate were estimated with an approach identical to 1), the only difference being the use of inverted Savitsky-Golay filter to enable detection of a drop in firing rate rather than an increase. 3) Latencies of cells that had very low spontaneous activity and which responded to stimulation with a membrane depolarisation ridden by one or few single spikes, were estimated based on the pooled original spike-trains and not aligned. Spikes denoting a response were mostly extremely well timed. An additional alignment usually introduced faulty shifts as a consequence the generally low spiking activity. The response latency was defined as the peak-time of the rate, which resembled the spike peak time in these conditions. Rate was estimated as detailed above. Independent of the method, the shortest mean latency amongst all latencies corresponding to neurons recorded in the same laboratory and set-up was subtracted from the remaining neurons latencies. This was done, so as to compensate for set-up specific differences in odour delivery time that introduce an artificial shift into the neurons latency.

Fluctuations in the membrane potential (Baseline Power, Stimulus Power).

After removal of spikes (cp. above), the average signal power ($P = 1/T \int_T^0 |s(t)|^2 dt$) over the baseline condition (500 ms pre-stimulus onset) and the baseline corrected power during stimulus presentation were calculated for each trial. Subsequently, signal power was averaged over all trials. To obtain stimulus related changes in signal power, mean baseline power was subtracted from the mean power within 500 ms post stimulus onset.

Area of stimulus related membrane potential deflections (Depolarisation / Hyper-polarisation).

The signal was smoothed using a Gaussian filter (width 25 ms). Deflections were defined as those parts of the signal where a threshold of the baseline voltage +/- two standard deviations was crossed and the signal maintained beyond threshold for a minimum of 250 ms without interruptions longer than 100 ms. If this incidence happened more than once throughout stimulus-duration, the area corresponding to the longest interval was calculated. Area values describing de- and hyper-polarisation were calculated for each trial of a given cell. From the retrieved values, positive and negative extreme was chosen to characterise the cell.

4.2.5 Statistical analysis.

Collecting descriptive values to characterise evoked activity results in a multi-dimensional data space. Moreover, some of the descriptors derive in part from the same origin and may hence be correlated or even carry redundant information. Principal Component Analysis (PCA) allows to reduce a set of possibly correlated variables into a smaller set of uncorrelated variables called Principal Components (PC) (Pearson 1901) that still retain the major information content. Using PCA in the present dataset allowed to reduce eleven descriptors to the first five PCs. Those PCs were sufficient to explain 80% of the underlying variance. Since the original variables differ in the scale on which observations was made, data were normalised using z-scores before it was subjected to the PCA algorithm.
To explore possible grouping of neurons according to the PCs of their evoked activity characteristics, unsupervised clustering using Ward linkage with Euclidean distances was performed. The incre-

mental method aims to reduce the variance within a cluster by merging data points into groups in a way that their combination gives the least possible increase in the within-group sum of squares (Ward 1963). The sum of squares as the distance measure (d) between two groups (r,s) is defined as $d(r,s) = \sqrt{2n_r n_s / n_r + n_s} \|\bar{x}_r - \bar{x}_s\|_2$, where $\|\;\|_2$ is the Euclidean distance, $\bar{x}$ the centroid of a cluster and n the number of elements in the cluster.

To derive a number of clusters interesting to interpretation, plotting the average within-cluster distance against the number of clusters offers a classical geometrical method (Thorndike 1953). The final number of clusters is determined where the curve markedly flattens, that is where the within cluster distance maximally decreases.

To make visible, whether clustering performed on simplified PC input would yield conclusive clusters for practical usage, I described neuron groups in the context of the actual data, that is descriptors of activity.

Data pre-processing, descriptor extraction and statistical analysis were performed using custom written routines in Matlab (2007a, The Mathworks Inc) and algorithms provided by the Matlab Statistics Toolbox.

4.3 Results

The aim of the present work was to find groups of AL neurons based on electro-physiological characteristics of evoked activity. For this purpose I defined a set of quantitative descriptors. Spontaneous rate, CV and regularity, were determined to characterise ongoing spiking activity. Stimulus related rate in- and decrease, latency and stimulus dependent latency variance were used as measures for evoked spiking activity. In order to include properties of sub-threshold activity spikes were removed from the signal. Baseline power was used to describe ongoing sub-threshold activity. Stimulus power and de- and hyperpolarisation of the spikeless signal served as measures for stimulus-related changes. Descriptive values were collected from odour responses of 67 AL neurons, for 25 of which the morphology was known. To omit information redundancy as a consequence of correlation between descriptors, reduce dimensionality, and diminish unspecific variance, data were subjected to PCA. Grouping of activity profiles was investigated by means of hierarchical clustering on the number of PCs, which explained a reasonable amount of variability. Group characteristics were evaluated by referring back to the descriptive values.

4.3.1 Clustering of AL neuron activity patterns based on spiking and sub-threshold information.

Transforming descriptors of spiking as well as sub-threshold activity by means of PCA, the first five PCs explained 80% of the underlying data variance. Sub-threshold descriptors gave major contribution to the first PC. In the second PC, descriptors of firing rate were of strongest influence. PCs three

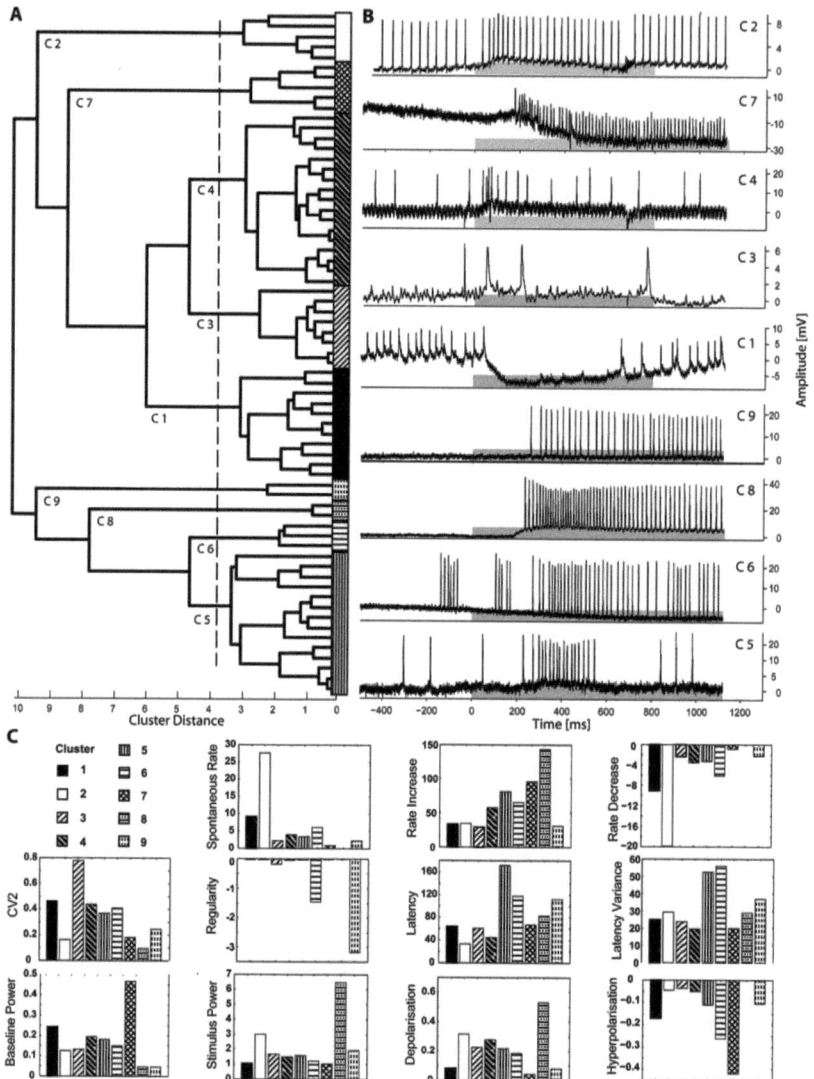

Figure 4.1: Cluster analysis based on spiking and sub-threshold criteria suggests nine conclusive groups. A) Cluster tree illustrating relationship within and between electro-physiological groups. The dotted line indicates the cut-off point defining the number of clusters. **B)** Exemplary traces of evoked activity for neurons from each cluster. Grey bars indicate stimulus onset and duration. **C)** Bar-plots give mean descriptive values for neuron groups allocated to each of the nine clusters.

to five combined measures of spiking activity in different combinations. On the basis of the geometrical stopping rule (cp. Methods) nine clusters (C1-9) were suggested (Fig. 4.1 A)). These nine groups were conclusive, not only in PC, but also in real data space. Accordingly, it was possible to outline

their distinguishing characteristics based on the defined set of descriptive values.
In C 1 (Fig. 4.1B/C: C1, black), eleven neurons were grouped together. All of these expressed pronounced, but temporally irregular, spontaneous firing in a characteristic combination with high baseline power. Neurons in this cluster responded with both, inhibition and excitation. Response onset was marked by slight firing rate in- or decrease of medium latency. Rate increase was often accompanied by hyperpolarisation. Two of the neurons in this cluster were putative LPNs, one was a confirmed mPN, a fourth a confirmed mlPN.

Table 4.1: Spiking and sub-threshold properties of nine groups of AL neurons.

Cluster	1	2	3	4	5	6	7	8	9
Spontaneous Rate [Hz]	9.2±3	27.7±10	2.3±1	3.5±2	3.4±4	6.1±2	0.86±1	0	2.3±3
Rate Increase [Hz]	33.6±20	34.5±17	28.9±6	57.7±33	81±33	65.3±14	96±39	143.6±6	30.9±4
Rate Decrease [Hz]	-9.1±3	-19.9±5	-2.3±1.2	-3.5±2	-3.3±4	-61±2	-0.08±1	0	-2.6±3
CV	.46±.1	.16±.1	.77±.08	.43±.18	.37±.21	.41±.03	.17±.05	.09±.03	.24±.23
Regularity	.02±.02	.02±.02	.17±.1	.05±.06	.03±.05	1.46±.37	.02±.01	.01±.01	3.18±.3
Latency [ms]	63±35	33±30	60±32	45±38	172±77	118±27	67±56	83±1	112±40
Latency Variability [ms]	24±13	29±22	24±8	19±9	53±20	56±5	20±15	29±10	37±11
Baseline Power [μV^2]	.24±.09	.12±.04	.13±.06	.19±.07	.18±.07	.15±.1	.47±.14	.05±.02	.05
Stimulus Power [μV^2]	1.05±.29	3.02±1.45	1.67±.51	1.49±.24	1.58±.75	1.21±.05	1.02±.38	6.48±1.66	1.92±1.31
Depolarisation [μV]	.08±.1	.32±.03	.23±.1	.28±.08	.22±.08	.18±.09	.04±.09	.53±.02	.08±0.11
Hyperpolarisation [μV]	-.17±.2	-.04±.1	-.03±.1	-.05±.1	-.11±.12	-.26±.2	-.42±.09	0	-.1±0.15
Morphology	1 mPN 2 IPN 1 mlPN		1 LN	1 LN	2 mPN 6 LN	8 mPN 2 IPN 3 LN	1 mPN 1 IPN 1 LN	4 mPN 1 IPN	2 mPN	2 mPN

The five neurons in C 2 (Fig. 4.1B/C C2, white), were marked by high, temporally regular spontaneous firing. In these neurons, sub-threshold activity was largely absent in baseline condition, but the typically fast responses during which firing rate increased slightly were accompanied by depolarisation of the membrane. The only neuron in this cluster for which morphological information was available was a confirmed LN.

In contrast, the eight neurons assembled in C 3 (Fig. 4.1B/C C3, orthogonal rising lines) were characterised by low spontaneous firing with very variable ISIs. Responses were in the lower medium range and characterised by depolarisation accompanied by single spikes rather than rate increase. Like in C 2 morphological information was present only for one cell, which again was a confirmed LN.

C 4 (Fig. 4.1B/C C4, orthogonal falling lines)) was, with seventeen neurons, the largest and most heterogeneous cluster. Representative group means for this cluster score close to average for almost all descriptors (Fig. 4.1C, Tab. 4.1). Responses of these neurons were characterised by rate increase with simultaneous membrane depolarisation and typically had a short latency. Morphological information was available for eight of the allocated neurons. Three were putative LNs, three more confirmed LNs and two confirmed mPNs.

The second large cluster, C 5 (Fig. 4.1B/C C5, vertical lines), comprising fourteen neurons, seemed to share characteristics of evoked, rather than spontaneous, activity. Following a very long latency, responses were of tonic or phasic-tonic nature, often accompanied by small de- sometimes hyper-polarisation of the membrane. Response latency varied greatly between different odours. With two confirmed and six putative mPNs, this cluster seemed clearly dominated by one type of morphology. However, two neurons were confirmed lPNs, two further LNs.

Three neurons were allocated to C 6 (Fig. 4.1B/C C6, horizontal lines). The neurons spontaneous activity was moderate with temporally irregular spikes of varying amplitude heights. Responses were typically phasic-tonic increases in firing rate, accompanied by membrane hyper- rather than de-polarisation. Latencies were long and varied greatly between odours. The cluster comprised neurons of all three morphological groups differentiated in this study.

C 7 (Fig. 4.1B/C C7, checker board) consisted of five neurons that were special in their absence of spontaneous firing but high baseline power. Evoked activity was characterised by phasic-tonic rate increase with simultaneous marked hyperpolarisation of the membrane. Three of the five neurons were confirmed mPNs, one more a putative mPN, the last an lPN.

The last two clusters were both very small, with only two allocated cells. Neurons in C 8 (Fig. 4.1B/C C8, dashed-solid lines) were similar to those in C 7 in that they had no spontaneous spiking activity and responded to stimulation with a pronounced rate increase. In contrast to neurons in C 7, rate increase went along with depolarisation and high stimulus-related power. One of the neurons was a confirmed mPN, the other a putative. The two cells in C 9 (Fig. 4.1B/C C9, dashed lines) foremost common feature was the irregularity in amplitude height. Responses had rather long latencies and were moderate rate increases with little contribution of sub-threshold activity. Both of the neurons were confirmed mPNs.

In conclusion, the clustering resulted in conclusive groups sharing similar electro-physiological profiles. Judging by the variability of the representative descriptive values, some groups appeared more heterogeneous than others. Four of the clusters contained predominantly mPNs (C5, C7, C8, C9). Neurons in these four clusters were similar in their very low spontaneous rate and a late response onset after which the firing rate increased. No cluster contained exclusively lPNs and only one contained more lPNs than other neurons (C1). Three clusters may tentatively be attributed LN morphologies (C2, C3, C4). Most clusters contained neurons of more than one morphological class.

4.3.2 Clustering of AL neuron activity patterns based on spiking information alone.

In PCA on descriptors of spiking as well as sub-threshold activity, the latter made major contribution to the first PC only. Accordingly, sub-threshold activity is a source of large variation but correlates little with spiking activity. I was wondering how much clustering results would change if performed on descriptors of spiking activity alone.

Running PCA on the seven descriptors of spiking activity, already the first three PCs could explain 70% of the underlying data's variance. Here, the first PC reflected most of all measures of firing rate. The second PC received strongest contribution from response onset measures and spike time regularity. Finally, the third PC combined influence of spike time regularity, amplitude regularity and rate increase. Thus all seven descriptors were combined in the first three PCs. The geometrical stopping rule suggested seven clusters (Cs1-7).

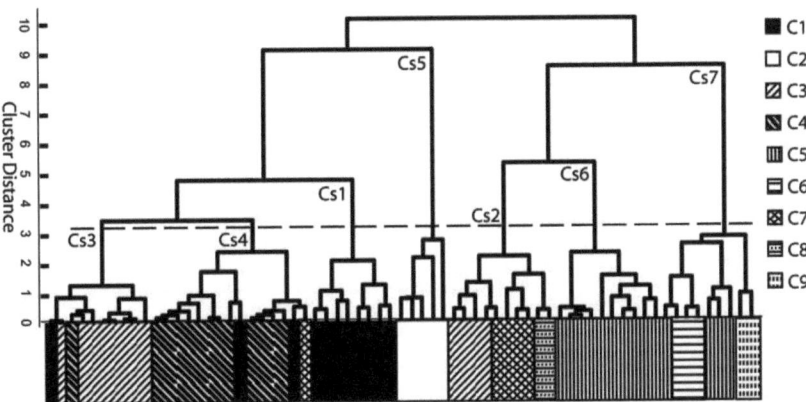

Figure 4.2: **Clustering based on spiking activity alone largely reproduces the grouping suggested by a combination of spiking and sub-threshold activity.** The cluster tree illustrates grouping of neurons based on descriptors of spiking activity. The dotted line gives the cluster distance cut-off point determining the number of spike based clusters (Cs). Pattern bars indicate to which of the nine clusters, which included sub-threshold information (C), each neuron had previously been allocated to.

Even though this clustering premised on less information than the previous attempt, grouping was largely preserved (Fig. 4.2). Neurons which were allocated to the same cluster (C) based on spiking and sub-threshold activity were likewise in a common spike based cluster (Cs). Neurons grouped in C4 were most scattered between different Cs clusters. This was not surprising, as C4 had been observed to comprise the most heterogeneous collection of activity profiles. Neurons in C2 and C3 seemed to be distinguished equally well through their spiking activity alone, as they were, again, without exception allocated to each a common cluster (Cs5, Cs3). Neurons in the small clusters C8 and C9 had singled out due to the characteristic membrane properties. These were now combined with other cells and incorperated into larger clusters.

The division and recombination of groupings effected the interrelationship between clusters. Neurons in C7 in particular were, judging from spiking activity only, classed much more alike with C8 and a subset of neurons from C4. C2, which was identical with Cs5, was now more closely related to neurons allocated to C1.

Taken together, leaving out on on sub-threshold information, grouping of neurons within clusters was largely preserved, while the interrelationship between clusters changed.

4.3.3 Distribution of PNs and LNs in the different cluster trees.

Cluster analysis of neurons based on combined spiking and sub-threshold information and on spiking information alone resulted in well comparable, conclusive groups. However, only in few cases could a morphological class be attributed to a certain electro-physiological profile. But, could on a broader scale, at least LNs and PNs be distinguished? To answer this question I considered the distribution of LNs and PNs between the two major branches of each of the two cluster trees.

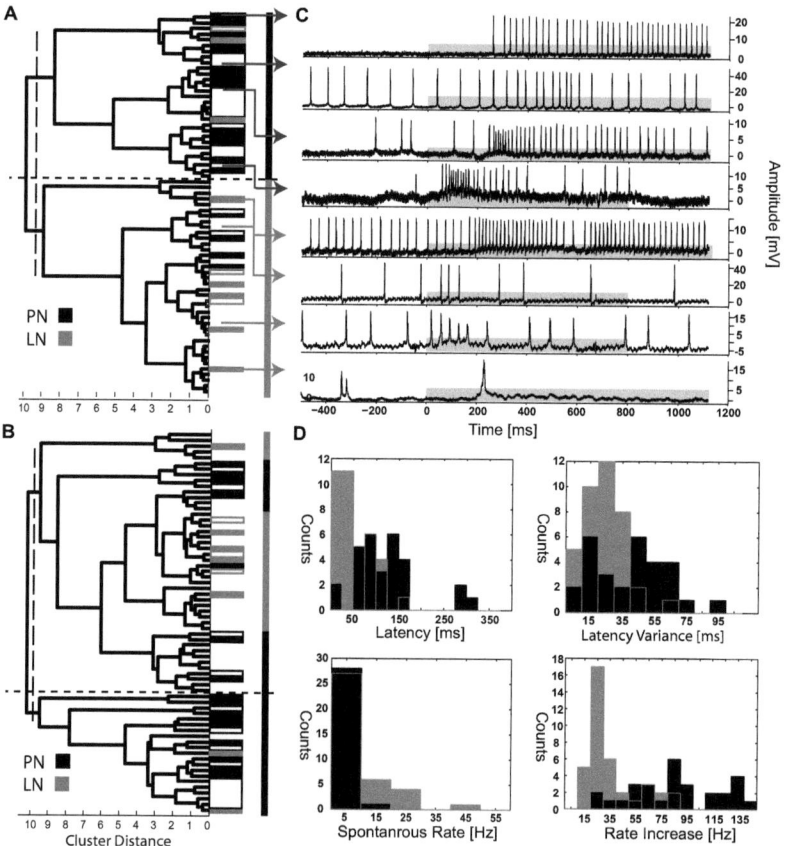

Figure 4.3: Comparison of LN and PN distribution between clusterings indicates better discrimination based on spiking activity alone. A) Distribution of PNs and LNs in groupings based on spiking activity. LNs are give in grey, PNs in black. Filled bars indicate confirmed morphologies, outlined bars putative morphologies. Vertical dotted lines indicate the cluster distance cut-off point, horizontal lines illustrate separation between branches. **B)** Distribution of PNs and LNs in groupings taking additional-sub threshold information into account. **C)** Exemplary traces of evoked activity from neurons in the upper PN-dominated cluster, and the lower LN-dominated cluster. **D)** Histograms illustrating differences in the range of selected descriptive values between neurons in the PN-dominated cluster (black) and LN-dominated cluster (LN).

4.4. Discussion

Table 4.2: Spiking properties of neurons in PN- and LN-dominated clusters.

	Spontaneous Rate [Hz]	Rate Increase [Hz]	Rate Decrease [Hz]	CV	Regularity	Latency [ms]	Latency Variability [ms]
LN-dominated	8±9	36.5±19	-7±6.4	.48±.2	.07±.09	52±38	24±12
PN-dominated	3.1±2	87.1±35	-3±3	.3±.8	.39±.89	124±75	40±22

None of the two trees segregated perfectly LNs from PNs (Fig. 4.3 A/B). In both solutions, one of the two groups was allocated clearly more PNs (86%, 18 of 21 for C_1; 25 of 29 for Cs_1) than LNs (10%, 2 of 21 for C_1; 3 of 29 for Cs_1). For the clustering which included sub-threshold information, the second group likewise contained more PNs (26%, 12 of 46 for C^2) than LNs (17%, 8 of 46 for C_2). In contrast, clustering based on descriptors of spiking properties alone yielded a second group with more LNs (19%, 7 of 37 for Cs_2) than PNs (16%, 6 of 37 for Cs_2). Under consideration of the few morphologically evident LNs contained in this study, this has to be seen as a valid separation between LNs and PNs.

Neurons in the LN-dominated cluster had, in the mean, higher spontaneous rates with temporally more irregular firing but regular amplitude heights as compared to neurons in the PN-dominated cluster. Evoked activity of neurons allocated to the LN-dominated cluster was, in the mean, of shorter, less variable latency, and marked by smaller increases in rate (Tab. 4.2C). However, looking at exemplary traces (Fig. 4.3C) and the distribution of selected descriptive values within the two groups (Fig. 4.3D), it becomes clear once more, that these are not homogeneous neuron clusters. Rather the reasonability to subdivide these in further groups is emphasised.

Obviously correlation between electro-physiological and morphological groups are not under simple constraints. Nevertheless, trends are visible. As such, PNs are better distinguished from LNs by properties of spiking activity alone.

4.4 Discussion

The aim of the present hand was to identify and characterise groups of AL neurons based on the electro-physiological properties extractable from patterns of evoked activity. In order to do so, quantitative descriptors of spiking and sub-threshold activity were defined and computed for 67 AL neurons. By using PCA, data was decorrelated and reduced. Groups were outlined by hierarchical clustering. On the foundation of spiking and sub-threshold activity, nine groups were distinguished. A simple one-to-one relationship between these nine different electro-physiological profiles and neuron morphology was not feasible. Still, clusters characterised by low spontaneous activity in combination with a long latency prior to a phasic-tonic response associated predominantly with mPNs. A second clustering, this time only on information of spiking activity, suggested less clusters but similar allocation of neurons. As a consequence of changed inter-relationship between clusters, PNs are better distinguished from LNs by properties of spiking activity alone.

4.4.1 Holistic or simplistic - how much information is necessary to distinguish meaningful clusters?

Electro-physiological characteristics of a neuron are not restricted to spiking properties. Sub-threshold activity similarly reflects the membranes ion-channels composition, which shapes the individual cell's active and passive membrane properties. Inhibition in particular is, in the absence of regular spontaneous firing, notoriously difficult to describe by spiking activity alone. Moreover, stimulus correlated depolarisation accompanied by spiking has previously been described as an incidental response pattern in AL neurons (Flanagan and Mercer 1989). I included measures of de- and hyperpolarisation, as well as baseline and stimulus correlated power as sub-threshold descriptors. In doing so, I hoped to achieve segregation of cells that showed inhibition as well as excitation from cells that only showed excitation on the one hand, and cells with response onset depolarisation from cells with generally vigorous sub-threshold activity on the other hand. Inferring from hierarchical clustering on the first five spiking-and-sub-threshold PCs, neurons with mixed inhibited and excited responses (C1) as well as neurons with onset depolarisation ridden by one or few spikes (C3) are confined by one group each (p. 4.3.1). Interestingly, these groups were quite well reproduced if the clustering was repeated on the first three PCs of spiking information and thus without taking sub-threshold information into account (C1 → Cs1, C3 → Cs3; Fig. 4.2). Instead, information reduction mostly affected segregation of neurons with little spontaneous firing and variations of phasic-tonic responses (C6, C7, C8, C9). The integration of these neurons into larger clusters can in part be explained by the loss of sub-threshold information. Neurons with pronounced stimulus correlated depolarisation (Fig. 4.2B; C8) were now clustered together with hyperpolarised neurons (Fig. 4.2B; C7). On the other hand, exclusion resulted in the reduction of noise. Measurements of sub-threshold activity by sharp electrodes is susceptible to artefacts as a consequence of rather unstable electrical contact (PING et al. 2008). Unstable contact can be reflected in slowly evolving polarisation (Fig. 4.2B; C6), or jumps in the baseline potential. Moreover, small shifts in membrane potential, which establish the true signal power, can be overshadowed by insufficiently shielded electrical noise, and thereby impact measurement of baseline power. Taken together, it appears that, even though measurements of sub-threshold information help to discern details of evoked activity, clustering based on PC transforms of spiking activity only is less impacted by noise and sufficient to produce conclusive groups.

On the assumption that descriptors of spiking activity alone are indeed sufficient to meaningfully discriminate between patterns of evoked activity, one might wonder whether the reductionistic detour via PCA is at all necessary. Applying PCA in the current case allowed to run clustering on three PCs instead of seven descriptive variables. Still, as group characteristics are largely reflected in the means of the genuine descriptors, it would be interesting to know in how far neuron groupings would differ if clustering was performed on the data directly. Grouping retrieved directly from measurable values would be desirable because these reflect the structure of the data more closely.

4.4.2 Science or fiction - may electro-physiological characteristics be used to predict AL-neuron morphology?

Morphologically distinct AL neurons have sometimes been attributed certain spiking patterns. Here I opted for the reversed approach and distinguished AL neurons on an electro-physiological foundation to attribute morphology.

Generally speaking, this approach did not yield a simple one-to-one relationship between physiology and morphology. Instead, groups of neurons with distinguishable electro-physiological characteristics seemed to correspond to morphologically similar neurons, the mPN class. A possible explanation can be found in the fact that mPNs, are even though morphological similar, not morphological homogeneous (Abel et al. 2001). Variations of phasic-tonic discharge patterns, as I found here to be clustered in different groups, could correspond to mPN subtypes.

A clearly more prominent morphological distinction can seen between the two large populations of m- and lPNs. Previous reports attribute complex response patterns of alternating excitation and inhibition to mPNs, while lPNs were found to have regular phasic-tonic responses (Abel et al. 2001; Müller et al. 2002). Elsewhere, this distinction is not confirmed, instead mPNs are reported to respond with mostly regular phasic-tonic excitation or simple inhibition, which is in agreement with the present results (Sun et al. 1993). A systematic difference between the two populations could not be found by means of hierarchical clustering, in the present data set. While the many clusters of mPNs seemed to have a common trend in their discharge behaviour, as pointed out above, the few lPNs included in the data set scattered broadly throughout different clusters. Only one of these qualified as a potential lPN cluster (C1/Cs1). Interestingly, this cluster was more closely related to electro-physiological characteristics of tentative LNs. The difficulty to associate lPNs and mPNs with distinct electro-physiological groups, calls into question whether the different clusters dominated by mPNs might truly be indicative of morphological subtypes. That a large population of mPNs shares long latencies and regular phasic-tonic response onset is, in regard to the fact that most clusters contained neurons of more than one morphological class, to be seen as a pronounced trend but not sufficient to predict neuron morphology.

Another pattern that has repeatedly been described (Flanagan and Mercer 1989) and illustrated (Galizia and Kimmerle 2004; Krofczik et al. 2009) in connection with hetero LNs is characterised by low, irregular spontaneous firing, short latencies and response inset depolarisation accompanied by few spikes. Hierarchical clustering yielded one rather homogeneous cluster with according properties (C3/Cs3). Only one of the neurons was morphologically documented, which was indeed a confirmed hetero LN. The all-in-all good agreement between literature and present findings proposes that the described discharge pattern can be associated with a sub-population of hetero LNs.

The most significant morphological distinction, surely, is that of PNs annd LNs. According to the results of this investigation, there seem to be prototypic activity patterns closely associated with mPNs as well as LNs. But can LNs and PNs be discerned based on their discharge patterns? This distinction is commonly assumed (Galizia and Kimmerle 2004; Krofczik et al. 2009) and, inferring from findings in mammals (McCormick et al. 1985), is likely to exist. Hierarchical clustering suggests that discharge patterns of LNs can not be perfectly segregated from PNs. However, under consideration of spiking activity descriptors alone a trend becomes apparent (cp. p. 48).

4.4.3 Utile or futile - why do we need established electro-physiological groups of AL neurons?

Obviously, comprehensive classification of AL neurons is easy undertaking. Definition of electro-physiological groupings which are neither to detailed nor to broad is a challenge in itself. Finding correlations between representative discharge patterns and neuron morphology most apparently defies to be described in a simple formula. Morphological classes on the other hand are well established and even though their common definition lacks details, they offer a framework based on which AL function can be speculated on. So, why do we need a classification based on electro-physiological descriptors at all?

The difficulty in attributing spiking patterns to morphologies in part originates in technical problems to collect both types of data for the same neuron. Not every recording lasts long enough to attempt subsequent staining of the cell, and not every attempt is successful. As long as the value lying in the electrical activity of the cell is only recognized in connection with its morphology, a lot of data has to remain either unused or undervalued. The establishment of electro-physiological distinct classes of AL neurons could facilitate the formation of research question that specifically targeted these neuron groups and thus increase the information that can be drawn from stand-alone electro-physiology: Do long latency, regular phasic-tonic AL neurons exhibit odour tuning? Are spontaneously active, short latency, neurons inhibited by mixtures? In computational works, electro-physiological classification could be of use as a further building block in modelling. Certainly, insights provided by such approaches will become even more informative if connected to neuron morphology. But, just as the function of a neuron arises from synergy between the independent properties - morphology, physiology and histochemistry -, so will comprehension of this function eventually be the result of synergy between meaningful and equal footed investigation of each of these properties.

CHAPTER 5

Conclusion and Outlook.

The main objective of this work was to characterise local interneurons (LNs) in the antennal lobe (AL) of the honey bee, firstly to identify functional LN sub-populations and secondly to describe physiological properties in which LNs differ from projection neurons (PNs).

In the first part of my investigations, I took a combined physiological-morphological approach to address these questions. I found that elemental coding of mixture components is performed by short latency neurons, probably LNs, as well as by long latency neurons, probably PNs. Configural coding of temporally imperfect mixtures appears to involve circuits different from elemental responses and requires more processing time. Morphological evidence further suggested that the two coding strategies are not necessarily accomplished by morphologically distinct neurons. I found, in particular, hetero LNs involved in both tasks. Instead elemental and configural coding might arise in the same neuron as a consequence of stimulus context and glomerular innervation pattern. A "multi-function" hetero LN acting in this fashion would require to receive sensory input from all its innervated glomeruli, sparse as well as dense.

Exploring the idea of impact connectivity has on functionality, I continued by studying morphological aspects of LNs. Assessing the spatial distribution of neurites within single and between different glomeruli, it appeared that LNs have spatial overlap with ORNs as well as PNs, but are more closely correlated to ORN defined structures of the AL. Many LN neurites form structures in the innermost part of the cap, emphasising a more fine-scaled compartmentalisation than commonly assumed. Detailed observations led me to differentiate six phenotypes. These phenotypes might be indicative of functional LN sub-populations and call into question the common notion that there are only two morphological groups of LNs.

The function of a neuron is inevitably the result of all its properties. Bearing this in mind, I opted in the last part of my investigations towards a classification of AL neurons on the basis of electrophysiology. Here I was not only looking for potential LN sub-populations, but also for the means to discriminate PNs from LNs without the need of additional morphological evidence. Clustering on principal components of descriptors of spiking activity, suggested conclusive groups of evoked activity patterns. However, only in some cases did neurons that exhibited similar evoked activity patterns belong to the same morphological group. Likewise, LNs and PNs did express tendencies to differ in terms of electrophysiology, but one should be cautious of using these as a predictor for neuron morphology.

I had started my investigation adopting the view that functional LN sub-populations do exist and help to implement important processing mechanisms in the AL. Reviewing the results of my investigations, I come to the conclusion that firm description of LN sub-populations is not trivial. Some of my results can be interpreted in favour of my adopted view. Others could likewise be used in support of the opposing view that denies the existence of functional sub-groups and suggest different properties of LNs to be only variations of one basic LN type. In the following I will elaborate on both of these ideas.

Odour processing in the AL - abolition of the hierarchy?

The olfactory system is considered a flat processing stream, because in theory, information can travel from ORNs to cognitive areas by crossing just a single OB/AL synapse. However, only looking at the OB's layered architecture it becomes clear that sequential and parallel processing does take place. In the AL, anatomical layers comparable to those in the OB are absent. However, physiological data provides strong evidence for the implementation of sequential and parallel circuits. Neurons that fulfil different tasks in different positions of the circuitry, can be considered functionally different sub-populations. Lets see what AL mechanisms we know of, put these in a relationship with what we have learned about LNs in this work and discuss in how far we end up having functional LN sub-populations.

Parallel pathways, it has been suggested, mediate gain control and contrast enhancement in the honey bee (Deisig et al. 2010; Galizia 2008; Sachse and Galizia 2002). It has been proposed that they correspond to morphologically distinct LNs. Homo LNs would mediate global gain control and hetero LNs glomerulus specific lateral inhibition. An underlying assumption based on which the two mechanisms are assigned to different groups of cells is that hetero LNs are polar neurons (Galizia 2008; Galizia and Kimmerle 2004). Homo LNs in contrast are speculated to be apolar neurons (Wilson and Mainen 2006). In this work I have reported findings which led me to suggest hetero LNs may be apolar, just like homo LNs. Previous investigations in moths have raised the possibility, that axonless LNs might posses active dendrite properties and comprise multiple functional subunits (Christensen et al. 1993), similar to the OB granule cell. Transferring these properties to hetero LNs it might be worthwhile to rethink the attribution of gain control and lateral inhibition to different LN sub-populations: In theory, hetero LNs could serve to implement both gain control and lateral inhibition. In this view, gain control could be achieved by reciprocal synapses in densely innervated glomeruli. Synapses from LN to ORN and vice versa were shown in cockroaches and are likely to exist in honey bees (Distler 1990; Gascuel and Masson 1991). Moreover, as shown in my work and elsewhere (Sun et al. 1993), LNs show large overlap with sensory neurons in their densely innervated glomerulus. Already weak ORN activation could trigger reciprocal inhibition, which would in fact be local, but as it acts equally on all activated ORNs, it would be difficult to separate from a global mechanism. Lateral inhibition could be the consequence of strong ORN or PN input in any of the innervated glomeruli. The synapses necessary for this type of connections are known to exist in cockroaches (Distler 1990; Malun 1991a,b). In this scenario LN activation arises simultaneously at spatially separated input sides, all of which are potential output sides. At which of the sites, inhibition eventually acts would depend on the balance between excitation and inhibition in each individual glomerulus. This mechanism would thus be highly odour specific and not necessarily symmetrical. This is in consistence with experimental as well as behavioural findings (review Galizia 2008). Pharmacological separability of lateral inhibition, acting on the level of PNs, and gain control, acting on the level of ORNs, that is mediated by similar types of LNs could be explained by action on different receptors at the site of the receiving synapse. Related mechanisms have been shown in drosophila (Wilson and Laurent 2005). To assess the potential that may lie in hetero LNs, it will be important to clarify whether these neurons are polar or not and if they possess active membrane properties.

It has been proposed that serial connection between LNs remove tonic inhibition from PNs (Christensen et al. 1993). Imagining a sequential network of two morphological distinct types of LNs, one would expect these neurons to meet certain criteria. One of these LN types should be well suited to receive ORN input and to connect to the second order LN. The second order LN in turn, should be well fitted to receive the first order LN input and connect to PNs. This layout results in two predictions: firstly, that LNs should cluster in groups of distinct latencies and secondly, that LNs should cluster in ORN correlated and PN correlated morphological groups. Consistently with earlier studies in the honey bee (Flanagan and Mercer 1989; Sun et al. 1993) I could not show groups of different latencies between LNs. In contrast, I found latencies within one LN could differ markedly depending on the stimulus context. Moreover, the potential of direct ORN input was given for each of the LNs I stained. The observations related in this work, suggest in fact an alternative model of serial connection between morphologically similar types of LNs. Again, these neurons would have to meet certain criteria. Both neurons should receive direct ORN input and give output to PNs. I found this to be the case for both homo and hetero LNs. Moreover, the two LNs should have overlapping but not identical glomerular innervation patterns. I found that neither hetero, nor homo LNs follow fixed innervation patterns or innervate all glomeruli such that again both types apply to this criterion. Each of the two neurons could, depending on the nature of the stimulus either be the first or second order LN. In any case the net outcome would be the integration of information from ORNs beyond the receptive field of the individual LN. This is a feature that may be useful in the important task of odour mixture processing.

Another serial connection which appears to follow a more strict hierarchy exists between LNs and PNs. LNs are found to respond with significantly shorter latency to odours than PNs (Krofczik et al. 2009). In accordance with this I found that latency is amongst the most useful descriptors in clustering of LNs and PNs. Consequently, PN activation might rely on LN activation. My findings generally support this notion, but small boosts of activity in PNs prior to inhibition provide indications for direct ORN to PN input. Findings in drosophila suggest that direct ORN input might even provide major contributions to the shaping of odour responses (Root et al. 2007). What importance does direct ORN to PN input have in the honey bee system? A first step towards answering this question would be to investigate the synaptic layout of the AL thoroughly. Modeling PN input and PN output from calcium imaging data in combination of sound standing ratio between ORN to PN versus LN to PN synapses could yield an approximation about the strength of the connections.

Summing up the ideas I have stated in this passage, the situation appears as follows: parallel processing pathways do exist in the honey bee. They may or may not be implemented by morphologically distinct neurons. Serial connections between LNs are likely to exist in the honey bee. They do not seem to follow a strict hierarchy and can employ morphologically similar or distinct LNs. Serial connections from LNs to PNs exist in the honey bee. Accordingly, sensory input passes LNs before it reaches PNs. However PNs may likewise receive direct sensory input, such that hierarchical order is again called into question. One might justifiably wonder, where to start searching for functional LN sub-populations in a system where seemingly every cell can take every other cell's place.

Finding function in LNs - looking for similarity amongst differences.

Much of the difficulty to outline LN sub-populations in the AL arises from the simple problem of spatial inseparability. The fact that interneurons in the OB distribute in layers has helped a lot to distinguish them as sub-populations with different properties on the one hand, and conclude about their function in the network on the other hand. For example, the identification of granule cells as reciprocal inhibitors of mitral cells was initially made based on the position of extracelluar recording electrodes in the layered tissue (Shepherd 1963). Similarly, in the AL properties and functions of ORNs and PNs are, in comparison to LNs, much better understood, because they form tracts and thus give rise to layer-equivalent structures. LNs in contrast are distributed throughout the AL and intermingle with neurites of both ORNs and PNs.

Intracellular methods allow to learn directly from a single LN about its individual properties. These approaches tackle the problem of spatial inseparability by assembling a collection of case studies. In this way it could be shown that LNs come in a large variety of morphologies. This is true, not only for the honey bee, but likewise for other frequently investigated insect (Chou et al. 2010; Christensen et al. 1993; Flanagan and Mercer 1989; Fonta et al. 1993; Matsumoto and Hildebrand 1981; Stocker et al. 1990; ?). A common finding of these studies is that some LNs show very symmetrical inter-glomerular innervation patterns that often include many glomeruli. Others have markedly asymmetric inter-glomerular innervation patterns. In the honey bee these have often been put at one level with homo and hetero LNs. However, as is apparent from the data I presented in this work, this is not necessarily true. Rather one could think that some interneurons are common to all species and hence ma be considere as basic elements of cortical circuits. These would be symmetric and asymmetric LNs. Other LNs are characteristic of particular species of insects and may reflect biologically relevant adaptations to the needs of the species. For the honey bee, these would be hetero LNs. In this light it appears a certainty that variations in morphology are not coincidences, but that there are sub-populations that have meaning. Comparative studies of LN development would be of great use to find out about which LNs may be considered basic elements and which not.

Just like there are similarities between different morphologies that allow to group LNs together, so are there similarities between electrophysiological profiles of different LNs that allow for conclusive grouping (Chou et al. 2010; Husch et al. 2009b). In this work I have shown that clustering on descriptors of spiking activity from AL neurons of the honey bee similarly produces groups with coherent profiles. Some of these activity patterns have repeatedly been shown or described in literature (Flanagan and Mercer 1989; Krofczik et al. 2009; Müller et al. 2002; Sun et al. 1993). For example the type of response I described as 'depolarisation ridden by one or a few spikes' appears to be so stable a pattern and is so frequently encountered that we should know more about its origin. However, electrophysiological properties and morphology are apparently not connected in simple one-to-one relationship. In order to find out about the system underlying the function of the neuron within the circuitry, we would have to identify the critical combination of characteristic parameters. These might be only a few amongst many. In order to understand which combinations are important, we will have to learn more about the single properties first. From intracellular methods we have learned how diverse LNs are, but is there a way to record systematically from the same population of neurons, preferably LNs to learn in what aspects they are similar?

Within a glomerulus, layers are partly represented in that ORNs do not invade the entire glomerulus (Galizia 2008). Due to this partial ORN innervation it is distinguished between two sub-structures in glomeruli of the honey bee: the ORN-defined cap and the inner core. In my study of LN morphology I found evidence for further compartmentalisation of glomeruli by LNs. Sparse, fork-like as well as dense, tree-like arbours of many LNs seem to accumulate in a ring-shaped intermediate layer. Other LNs invaded the entire glomerulus. These morphological characteristics are good criteria to differentiate between LN-populations. Moreover, they hint towards a layered, concentric architecture of glomeruli. Spatial distribution of neurotransmitters and -peptides supports the idea of more fine-scaled glomerular structure (Bornhauser and Meyer 1997; Dacks et al. 2010; Kreissl et al. 2010), and similar findings of structured arborisation of LNs have been reported for drosophila (Chou et al. 2010; Tanaka et al. 2009). However, drosophila LN arbours do not structure in layers but in patches. The concentric architecture of the honey bee glomerulus could pose a point of action to investigate the physiology of a sub-population of LNs. These would first of all be hetero LNs with fist-like dense arbours. Analogue to simultaneous recordings from layers of different depth in the neo-cortex, one could consider extracellular recording from single physiologically identified glomeruli with tetrodes of graded length. In the signal of a carefully placed tetrode, each of the individual tips would ideally be dominated by single unit responses in different glomerular layers. Provided that a suitable dye could be found, the approach might as well be suitable for line-scanning with high temporal resolution (Junek et al. 2010). Before such attempts are made, it will be important to collect more detailed morphological data about the different LN arborisations in glomeruli. In particular the localisation of sparse, fork-like arbours will have to be investigated. This does not necessarily involve single cell stainings. Bulk-staining in combination with sensory backfills should be sufficient to resolve the question.

Taken together the results of this work suggest that LNs are variable in many properties. As I have pointed out earlier, a cells function arises from all its properties. Neurons express a natural variety in their features such that no perfect correlation between features might be found. Accordingly, population of neurons may be considered functional when the variability of features within the population is much smaller than across populations. The implication of multi-functional hetero LNs, as I have discussed it above, suggest that despite of the seeming diversity, many of the tasks associated with the AL could be accomplished by a rather small set of fundamentally different LNs. In fact, one may argue that a brains capacity would grow with the number of copies of a complex neuron, rather than with the amount of different but simple neurons In this sense, an ultimate classification might be based on the role an LN plays in the circuit, like I have stated in the beginning of this discussion. However, until we have gained better understanding in the contribution each element of the AL networks makes, such an approach remains the ideal.

Summary

The antennal lobe (AL) is the primary olfacory center of the honey bee. It is the site of interaction between olfactory receptor neurons, and two types of AL neurons: local interneurons (LNs) that are restrained to the AL, and projection neurons (PNs) that relay output to higher processing areas. The present work investigates physiological and morphological properties of honey bee AL neurons, LNs in particular. The individual studies, summarized here, are united by the underlying attempt to infer potentially functional LN sub-populations from the described characteristics.

My first objective was to investigate how individual AL neurons encode a type of complex information content, the olfactory system is challenged with every day: an odour mixture (Chapter 2). I stimulated with mono-molecular odorants, their temporally perfect-,and imperfect binary mixture to reproduce a natural dynamic odour environment. Single cell's responses were recorded intracellularly.
Response patterns between different neurons varied in many details but were generally speaking either rate changes to excitation or inhibition, respectively, or membrane depolarisation accompanied by one or a few single spikes (cf.: 2.3.2).
Irrespective of its individual response pattern, each neuron, challenged with the binary mixtures, responded in one of two possible ways: elemental, or configural. About half of the neurons responded elementally, i.e. responses evoked by mixtures reflected the underlying feature information from one of the components. The other half exhibited configural responses, i.e. responses evoked by mixtures represented these as clearly different from their single components (cf.: 2.3.3).
A question immediately arising is, whether these two types of encoding could be associated with different sub-populations of neurons. Referring to the neuron's latencies as an indicator of position within the circuitry, I found that elemental neurons divided in earl responders and late responders whereas latencies of configural coding neurons concentrate in between these divisions (cf.: 2.3.4). From this finding one may infer that elemental odour coding is not confined to only one sub-population of neurons. In fact, it is more likely that LNs and PNs, which have previously been shown to differ significantly in latency, can both exhibit elemental coding (cf.: 2.4.2). Latencies of neurons with configural responses expressed a tendency to respond faster to single components than to imperfect mixtures. This finding made me think that these neurons may participate in multiple processing circuits.
Both of the above assumptions were confirmed by exemplary morphological data (cf.: 2.3.5). For each of the two groups of 'elemental neurons', early and late responders, I could obtain an exemplary staining. The early responding neuron was confirmed as an LN, while the late responding neuron was a PN. More surprisingly, however, I found that one of the 'configural neurons' was a hetero LN, just like the short latency elemental neuron. By comparing the inter-glomerular innervation patterns of the two hetero LNs with odour-specific glomerular activation maps, derived from calcium imaging, I hoped

to find an explanation for the difference in odour coding. Indeed, the elemental hetero LN innervated one of the responsive glomeruli densely. The configural hetero LN, in contrast, innervated glomeruli that were responsive to the chosen stimuli only sparsely. Based on the combined morphological and physiological evidence, I propose to consider hetero LNs as multi-function neurons. Multi-function neuron here means, the possibility to be recruited by different circuits such that elemental as well as configural odour-processing are performed by the individual neuron in a stimulus-context dependent manner (cf.: 2.4.3).

The proposed multi-function hetero LN would require the ability to receive sensory input and give output in both, sparsely and densely branching arbours. Are these requirements met on the local scale of individual intra-glomerular arborisation? And judging by the global inter-glomerular innervation, in how far are LNs generally tailored to receive and redistribute direct sensory input? Or are innervation patterns rather oriented towards PNs? These are questions I hoped to answer by analysis of LN morphologies under functional aspects (Chapter 3).
I reconstructed morphologies of single neurons from stainings of different LNs. Neuron reconstructions were transformed to fit into a common reference frame. By this means, it got possible to evaluate which of the four sensory tracts (afferent fields) and which of the two AL hemilobes (efferent field) received innervations. To analyse intra-glomerular arborisation I reconstructed cap, and where possible core of single glomeruli as well as sparse or dense arborisations of the LNs innervating them.
The inter-glomerular innervation pattern of LNs differed widely on an individual scale, but appeared to correlate with the division of afferent fields rather than efferent fields (cf.: 3.3.1). This observation suggests that at least a considerable sub-population of LNs is tailored to collect and integrate meaningfully related ORN input (cf.: 3.4.1). Investigating intra-glomerular arborisation I found markedly different branching patterns for sparse as well as dense arbours. All of these had the potential to establish contact with sensory neurons in the glomerular cap as well as PNs and other LNs in the core (cf.: 3.3.2). On these grounds, morphological requirements for a multi-functional hetero LN are fulfilled.
Having come so far it seems that LNs express a certain functional communality. Still, differences in local and global branching patterns, amongst other characteristics, are obvious. Continuing from a perspective of functional morphology, I assembled a toolbox of morphological descriptors based on which I differentiate six LN-phenotypes (cf.: 3.3.4). Clearly, under the objective of finding functional LN sub-populations the convenient division in only two broad groups of homo- and hetero LNs has to be reconsidered (cf.: 3.4.2).

Systematic differences in morphology, like those based on which I had distinguished six different LN Phenotypes, are indicators for functional differences between neurons. But, as the function of a neuron is inevitably the result of all its properties, the same is true for measures of differences in physiology. Classification based on spiking properties of single neurons has decidedly facilitated the investigation of inter-neurons in the mammalian neocortex. I wondered whether the activity patterns observed in the honey bee, by me and others, could likewise be separated in conclusive groups (Chapter 4).

To approach this question I analysed single cell recordings from a set of AL neurons which included different LNs, PNs and morphologically unidentified neurons. Collected descriptive values of spiking and sub-threshold activity that could be extracted from odour-evoked responses were decorrelated and reduced by means of PCA and subsequently clustered by means of an automated hierarchical algorithm.

Referring back to the original descriptive values, some of the suggested groups were immediately conclusive whereas others appeared more heterogeneous (cf.: 4.3.1). Repeating the clustering on PCs derived from spiking activity only, the allocation of neurons within clusters was largely preserved while the inter-relationship between groups changed (cf.: 4.3.2). On this ground I conclude that electro-physiological classification of honey bee AL-neurons is feasible and that the information contained in spiking activity alone is sufficient for this purpose (cf.: 4.4.1).

If it is possible to classify groups of AL neurons according to their discharge patterns, could the classification criteria be used to predict neuron morphology? Examining the distribution of morphological identified LNs, mPNs and lPNs between the suggested groups, I had to conclude that, there is no simple one-to-one relationship between electro-physiological profiles and neuron morphology (cf.: 4.3.3/4.3.1). Still, trends are present in that discharge patterns seem to be typically associated with mPNs and others are strongly suggested to be correlated with LNs (cf.: 4.4.2). As a whole, LNs and PNs are better distinguished based of spiking activity alone (cf.: 4.3.3).

Zusammenfassung

Der Antnnallobus (AL) ist das primäre Hirnareal für Geruchsverarbeitung der Honigbiene. Hier treffen die drei Neuronentypen mit grundlegend unterschiedlichen Aufgaben aufeinander: Die olfaktorischen Rezeptorneurone (ORN) liefern Informationen über die Düfte der Umgebung. Die lokalen Interneurone (LN) verschalten Regionen innerhalb des AL miteinander. Die Projektionsneurone (PN) schließlich senden die im AL aufgearbeitete Duftinformationen an weiterführende Verarbeitungsareale. LNs und PNs zusammen werden auch als AL-Neuronen bezeichnet. In der vorliegenden Arbeit habe ich physiologische und morphologische Eigenschaften von AL-Neuronen, insbesondere der LNs, untersucht. Allen hier zusammengefassten Studien liegt als gemeinsamer Gedanke zu Grunde, die beschriebenen Charakteristika zu nutzen um potentiell funktionelle Subpopulationen von LNs zu identifizieren.

Mein erstes Interesse galt der Frage, wie individuelle AL-Neurone die komplexe Information kodieren, die in einer Duftmischung enthalten ist (Kapitel 2). Um die Verhältnisse in einem natürlichen, und damit dynamischen Geruchsumfeldes nach zu empfinden, habe ich zwei mono-molekulare Düfte, so wie deren zeitlich perfekte und imperfekte Mischungen als Stimuli verwendet. Die Duftantworten der einzelnen Neurone wurden mittels intrazellulärer Ableitungen gemessen.
Antwortmuster unterschieden sich zwischen den einzelnen Neuronen in vielen Aspekten, konnten aber grob verallgemeinert als entweder erregte, bzw. unterdrückte Veränderungen der Feuerrate, oder Depolarisation der Membran begleitet von vereinzelten Aktionspotentialen beschrieben werden (vgl.: 2.3.2).
Unabhängig vom Antwortmuster, reagierte jedes Neuron auf Stimulation mit binärere Duftmischung in einer von zwei möglichen Weisen: elementar oder konfigural. Ungefähr die Hälfte der abgeleiteten Zellen kodierte Duftmischungen elementar, dass heißt die Antwort spiegelte Informationen über eine der zu Grunde liegenden Einzelkomponenten wieder. Die andere Hälfte kodierte die Duftmischungen konfigural, dass heißt Antworten auf Duftmischungen waren deutlich unterschiedlich von denen der Einzelkomponenten (vgl.: 2.3.3).
Eine Frage die sich bei dieser Beobachtung fast automatisch stellt ist, ob diese Beiden unterschiedlichen Weisen der Kodierung von zwei Subpopulationen von Neuronen herrühren. Bei der Betrachtung der Latenzzeiten, die Rückschlüsse auf die Position des jeweiligen Neurons im Verarbeitungskreislauf zulassen, zeigte sich, dass elementar kodierende Neurone in jeweils eine früh- und eine spätantwortende Untergruppen zerfallen. Die Latenzen der konfigural kodierenden Neurone fielen genau zwischen diese zwei Gruppen (vgl.: 2.3.4). Dieser Befund gibt Anlaßanzunehmen, dass elementare Kodierung nicht einem einzelnen Typ Neuron zuzuweisen ist. Tatsächlich ist es wahrscheinlicher, dass sowohl Neurone vom Typ der LNs als auch solche vom Typ der PNs, die sich deutlich in ihren Latenzzeiten unterscheiden, elementare Duftantworten zeigen (vgl.: 2.4.2). Für die konfigural kodierenden Neurone war eine solche Zweiteilung er Latenzen nicht zu sehen. Hier war jedoch

ein anderer Trend erkenntlich, nämlich der, schneller auf Einzelkomponenten als auf imperfekte Mischungen zu antworten. Diese Beobachtung deutete ich als Hinweis auf eine mögliche Einbindung diese Neuron in mehrere, parallele Verarbeitungswege.

Die Annahmen, die ich gegenüber beiden Gruppen, konfiguralen und elementar kodierenden Neuronen, gemacht hatte, wurden bestärkt durch exemplarische Färbungen abgeleiteter Neurone (vgl.: 2.3.5). Für jede der zwei elementaren Untergruppen, konnte jeweils eine Färbung erhalten werden. Das früh-antwortende Neuron war in der Tat ein LN, das spät-antwortende ein PN. Überraschender hingegen war die Entdeckung, das ein erfolgreich gefärbtes configural kodierendes Neuron, zur gleichen morphologischen Gruppe gehörte, wie das früh-antwortende elementare Neuron, zu den hetero-LNs. Wie kam es dazu das morphologisch ähnliche Neurone Mischungen so unterschiedlich kodieren? Ich hoffte eine Erklärung zu finden, in dem ich die inter-glomerulären Innervationsmuster der individuellen Neurone mit den duftspezifischen glomerulären Aktivitätsmustern, wie sie durch Kalziumbildgebung bekannt sind, verglich. Anhand dieses Vergleiches konnte ich feststellen, dass des elementar antwortende hetero-LN einen der auf den Stimulus ansprechenden Glomeruli stark innervierte. Das Neuron, das die Duftmischung konfigural beantwortet hatte, innervierte hingegen Glomeruli die auf den Duft ansprechen nur schwach. Auf Grund der kombinierten physiologischen und morphologischen Befunde gehe ich davon aus, dass hetero-LNs als multi-funktions Neurone betrachtet werden können. Multi-funktionalität entspricht in diesem Fall der Möglichkeit in über unterschiedliche Verarbeidungspfade angesprochen zu werden, wodurch elementare und konfigurale Kodierung in Abhängigkeit vom Stimuluskontext durch das gleiche Neuron geleistet werden können (vgl.: 2.4.3).

Ein solches multi-funktionales hetero LN setzt voraus, dass sensorischer Eingang aus stark innervierten, aber auch aus schwach innervierten Glomeruli erhalten wird. Ist diese Voraussetzung auf der lokalen Ebene von individueller intra-glomerulärer Verästelung erfüllt? Und, in Hinblick auf die globale, inter-glomeruläre Verzweigung, wäre LN-Morphologie auf die Aufgabe zugeschnitten, sensorische Information zu empfangen und um zu verteilen? Oder sind die Verzweigungsmuster eher mit Rücksicht auf PN gegebene Strukturen orientiert? (Kapitel 3).
Ausgehend von Einzelzellfärbungen, rekonstruierte ich die Morphologie unterschiedlicher LN. Die rekonstruierten Neurone wurden durch Transformation in einen gemeinsamen Bezugs-AL, gesetzt. Anhand dieses Bezug-ALs war es nun möglich abzuschätzen, welche der vier sensorischen Trakte (afferente Felder) und welche der zwei AL Hemiloben (efferente Felder) vom individuellen Neuron innerviert wurden. Um die intra-glomeruläre Verästelung zu untersuchen rekonstruierte ich zusätzlich zu den Neuriten der LNs, die Kappe und wo möglich auch den Kern des innervierten Glomerulus selber.
Das Muster der inter-glomerulären Verzweigungen war zwischen LNs sehr unterschiedlich, korrelierte aber besser mit der Unterteilung der afferenten Felder als mit der der efferenten Felder (vgl.: 3.3.1). Folglich kann davon ausgegangen werden, das zumindest ein beträchtlicher Anteil an LNs dafür ausgelegt ist, Informationen von ORNs zu empfangen und zwischen ihnen zu integrieren (vgl.: 3.4.1). Bei Betrachtung der intra-glomerulären Verästelungen, stellte ich fest, das stark

verzweigte, wie auch schwach verzweigte Neurite in systematische Untergruppen differenziert werden konnten. Jedes dieser Verästelungsmuster zeigte räumliche Überlappung mit der Kappe und dem Kern des Glomerulus (vgl.: 3.3.2). Offensichtlich teilen LNs gewisse gemeinschaftliche, funktionelle Eigenschaften. Nichts desto trotz sind sie in vielerlei Hinsicht deutlich voneinander unterschiedlich. Weiterhin ausgehend von einem funktionellen Blickwinkel auf morphologische Charakteristika, definierte ich eine Anzahl von systematisch unterscheidbaren Eigenschaften, anhand derer sich sechs LN-Phenotypen ergaben (vgl.: 3.3.4). Damit wird deutlich, dass die Einteilung von LNs in nur zwei Subpopulationen, nämlich hetero und homo LNs überdacht werden muss (vgl.: 3.4.2).

Systematische morphologische Unterschiede sind Hinweise darauf, das die betreffenden Neurone möglicherweise unterschiedliche Funktinen erfüllen. Da die Funktion eines Neurons aber unbestritten das Ergebnis des Zusammenspiels aller seiner Eigenschaften ist, gilt das Selbe für physiologische Unterschiede. So hat die Klassifizierung von Einzelzellen aufgrund ihres Feuerprofils die Erforschung von etwa neokortikalen Interneuronen entschieden vorangetrieben. Inspiriert durch Arbeiten im Neokortex, wollte ich wissen, ob die unterschiedlichen Antwortmuster, die im AL der Honigbiene beobachtet werden, in ähnlicher Weise aussagekräftig Unterteilt werden können (Kapitel 4).
Ich verfolgte diesen Ansatz, indem ich Einzelzellableitungen von unterschiedlichen LNs, PNs sowie morphologisch nicht identifizierter Neurone analysierte. Deskriptive Werte des Feuerverhaltens, aber auch für Veränderungen des Membranpotentials, wurden aus Aufnahmen von duftevozierter Aktivität erhoben und anschließend mit Hilfe von PCA dekorreliert und reduziert. Die so transformierten Werte wurden unter Verwendung eines hierarchischen Clusteralgorythmus in Ähnlichkeitsgruppen eingeteilt.
Wurden die so erhaltenen Gruppen von Neuronen anhand der ursprünglichen Deskriptoren charakterisiert, zeigte sich das einige der Gruppen fast stereotype Gemeinsamkeiten aufweisen, während andere deutlich heterogener sind (vgl.: 4.3.1). Wurde das Clustering auf Hauptkomponenten die ausschließlich Deskriptoren von Feuerverhalten berücksichtigten wiederholt, veränderte sich die Zuteilung von Neuronen innerhalb einer Gruppe kaum. Die Verwandtschaftsbeziehungen zwischen Gruppen hingegen waren verändert (vgl.: 4.3.2). Anhand dieses Ergebnisses wird deutlich, dass es möglich ist AL-Neurone der Honigbiene anhand ihrer Elektrophysiologie zu klassifizieren und das zu diesem Zweck die im Feuerverhalten enthaltene Information hinreichend ist (vgl.: 4.4.1).
Wenn es möglich ist AL-Neurone anhand ihrer Aktivitätsmuster zu unterscheiden, können die entsprechenden Unterscheidungskriterien genutzt werden um eine Vorhersage über die Morphologie des Neurons zu machen? Betrachtet man zu diesem Zweck die Verteilung von morphologische identifizierten LNs, mPNs und lPNs zwischen den elektrophysiologischen Gruppierungen, muss festgestellt werden, dass eine eindeutige Zuweisung nicht möglich ist (vgl.: 4.3.3/4.3.1). Nichts desto trotz sind deutliche Trends zu erkennen. So sind einige Feuermuster typischerweise mit mPNs verknüpft, während andere aller Wahrscheinlichkeit nach von LNs stammen (vgl.: 4.4.2). Im Großen und Ganzen, sind

PNs besser von LNs auf der Grundlage ihrer Feuermuster und ohne zusätzliche Beschreibung des Membranpotentials zu unterscheiden (vgl.: 4.3.3).

Danksagung

Mein Dank gilt Prof. Dr. C. Giovanni Galizia, Prof. Dr. Martin P. Nawrot, Christoph J. Kleineidamm, meinen Eltern und Dr. Cornelia Eisenach.

Bibliography

Abel, R. (1997). *Das olfaktorische Systemm der Honigbiene. Elektrophysiologische und morphologische Charakterisierung von Antennallobus Neuronen und deren Beteiligung beim olfaktorischen Lernen*. PhD thesis, Berlin: Freie Universität.

Abel, R., Rybak, J., and Menzel, R. (2001). Structure and response patterns of olfactory interneurons in the honeybee, Apis mellifera. *J Comp Neurol*, 437(3):363–383.

Aungst, J. L., Heyward, P. M., Puche, A. C., Karnup, S. V., Hayar, A., Szabo, G., and Shipley, M. T. (2003). Centre-surround inhibition among olfactory bulb glomeruli. *Nature*, 426(6967):623–629.

Balderrama, N., ñez, J. N., Giurfa, M., Torrealba, J., Albornoz, E. D., and Almeida, L. O. (1996). A detereent response in honeybee (apis mellifera) foragers: Dependence on disturbance and season. *J Insect Physiol*, 42:463–470.

Balderrama, N., ñez, J. N., Guerrieri, F., and Giurfa, M. (2002). Different functions of two alarm substances in the honeybee. *J Comp Physiol A Neuroethol Sens Neural Behav Physiol*, 188(6):485–491.

Baraldi, R., Rapparani, F., Rossi, F., Latella, A., and Ciccioli, P. (1999). Volatile organic compound emissions from flowers of the most occuring and economically important species of fruit trees. *Phys. Cem. Earth*, 24:729–732.

Barbara, G. S., Zube, C., Rybak, J., Gauthier, M., and GrÃ¼newald, B. (2005). Acetylcholine, GABA and glutamate induce ionic currents in cultured antennal lobe neurons of the honeybee, Apis mellifera. *J Comp Physiol A Neuroethol Sens Neural Behav Physiol*, 191(9):823–836.

Berg, B. G., Schachtner, J., Utz, S., and Homberg, U. (2007). Distribution of neuropeptides in the primary olfactory center of the heliothine moth Heliothis virescens. *Cell Tissue Res*, 327(2):385–398.

Bicker, G. (1999). Histochemistry of classical neurotransmitters in antennal lobes and mushroom bodies of the honeybee. *Microsc Res Tech*, 45(3):174–183.

Bierfeld, J. (2009). *Characterization of immunhistochemically identified local interneurons in the antennal lobe of the honeybee*. Master's thesis, Universität Konstanz.

Bierfeld, J., Charlina, N., Galizia, G., and Kreissl, S. (2011). Immunostaining reveals at least six subpopulations of olfactory local interneurons contributing to the apis mellifera antennal lobe network. In *9th Göttingen Meeting of the German Neuroscience Society*.

Bornhauser, B. C. and Meyer, E. P. (1997). Histamine-like immunoreactivity in the visual system and brain of an orthopteran and a hymenopteran insect. *Cell Tissue Res*, 287(1):211–221.

Carlsson, M. A., Diesner, M., Schachtner, J., and Nässel, D. R. (2010). Multiple neuropeptides in the drosophila antennal lobe suggest complex modulatory circuits. *J Comp Neurol*, 518(16):3359–3380.

Carlsson, M. A., Galizia, C. G., and Hansson, B. S. (2002). Spatial representation of odours in the antennal lobe of the moth spodoptera littoralis (lepidoptera: Noctuidae). *Chem Senses*, 27(3):231–244.

Carlsson, M. A. and Hansson, B. S. (2003). Dose-response characteristics of glomerular activity in the moth antennal lobe. *Chem Senses*, 28(4):269–278.

Chou, Y.-H., Spletter, M. L., Yaksi, E., Leong, J. C. S., Wilson, R. I., and Luo, L. (2010). Diversity and wiring variability of olfactory local interneurons in the drosophila antennal lobe. *Nat Neurosci*. Journal club 15.02.2010.

Christensen, T. A., D'Alessandro, G., Lega, J., and Hildebrand, J. G. (2001). Morphometric modeling of olfactory circuits in the insect antennal lobe: I. Simulations of spiking local interneurons. *Biosystems*, 61(2-3):143–153.

Christensen, T. A., Waldrop, B. R., Harrow, I. D., and Hildebrand, J. G. (1993). Local interneurons and information processing in the olfactory glomeruli of the moth Manduca sexta. *J Comp Physiol [A]*, 173(4):385–399.

Christensen, T. A., Waldrop, B. R., and Hildebrand, J. G. (1998a). Gabaergic mechanisms that shape the temporal response to odors in moth olfactory projection neurons. *Ann N Y Acad Sci*, 855:475–481.

Christensen, T. A., Waldrop, B. R., and Hildebrand, J. G. (1998b). Multitasking in the olfactory system: context-dependent responses to odors reveal dual gaba-regulated coding mechanisms in single olfactory projection neurons. *J Neurosci*, 18(15):5999–6008.

Connors, B. W. and Gutnick, M. J. (1990). Intrinsic firing patterns of diverse neocortical neurons. *Trends Neurosci*, 13(3):99–104.

Dacks, A. M., Reisenman, C. E., Paulk, A. C., and Nighorn, A. J. (2010). Histamine-immunoreactive local neurons in the antennal lobes of the hymenoptera. *J Comp Neurol*, 518(15):2917–2933.

Deisig, N., Giurfa, M., Lachnit, H., and Sandoz, J.-C. (2006). Neural representation of olfactory mixtures in the honeybee antennal lobe. *Eur J Neurosci*, 24(4):1161–1174.

Deisig, N., Giurfa, M., and Sandoz, J. C. (2010). Antennal lobe processing increases separability of odor mixture representations in the honeybee. *J Neurophysiol*, 103(4):2185–2194.

Devaud, J. M. and Masson, C. (1999). Dendritic pattern development of the honeybee antennal lobe neurons: a laser scanning confocal microscopic study. *J Neurobiol*, 39(4):461–474.

Distler, P. (1990). Gaba-immunohistochemistry as a label for identifying types of local interneurons and their synaptic contacts in the antennal lobes of the american cockroach. *Histochemistry*, 93(6):617–626.

Distler, P. G., Gruber, C., and Boeckh, J. (1998). Synaptic connections between GABA-immunoreactive neurons and uniglomerular projection neurons within the antennal lobe of the cockroach, Periplaneta americana. *Synapse*, 29(1):1–13.

Flanagan, D. and Mercer, A. R. (1989). Morphology and response characteristics of neurones in the deutocerebrum of the brain in the honeybeeapis mellifera. *J Comp Physiol*, 164:483–494.

Fonta, C., Sun, X.-J., and Masson, C. (1993). Morphology and spatial distribution of bee antennal lobe interneurones to odours. *Chem. Senses*, 18(2):101–119.

Friedrich, R. W. and Korsching, S. I. (1997). Combinatorial and chemotopic odorant coding in the zebrafish olfactory bulb visualized by optical imaging. *Neuron*, 18(5):737–752.

Galizia, C. G. (2008). *The Senses: A Comprehensive Reference*, chapter Insect Olfaction, pages 725–769.

Galizia, C. G., Joerges, J., Küttner, A., Faber, T., and Menzel, R. (1997). A semi-in-vivo preparation for optical recording of the insect brain. *J Neurosci Methods*, 76(1):61–69.

Galizia, C. G. and Kimmerle, B. (2004). Physiological and morphological characterization of honeybee olfactory neurons combining electrophysiology, calcium imaging and confocal microscopy. *J Comp Physiol A Neuroethol Sens Neural Behav Physiol*, 190(1):21–38.

Galizia, C. G., McIlwrath, S. L., and Menzel, R. (1999). A digital three-dimensional atlas of the honeybee antennal lobe based on optical sections acquired by confocal microscopy. *Cell Tissue Res*, 295(3):383–394.

Galizia, C. G. and Menzel, R. (2000). Odour perception in honeybees: coding information in glomerular patterns. *Curr Opin Neurobiol*, 10(4):504–510.

Bibliography

Galizia, C. G. and Rössler, W. (2010). Parallel olfactory systems in insects: anatomy and function. *Annu Rev Entomol*, 55:399–420.

Gascuel, J. and Masson, C. (1991). A quantitative ultrastructural study of the honeybee antennal lobe. *Tissue Cell*, 23(3):341–355.

Gottfried, J. A. (2010). Central mechanisms of odour object perception. *Nat Rev Neurosci*, 11(9):628–641.

Hayar, A., Karnup, S., Ennis, M., and Shipley, M. T. (2004). External tufted cells: a major excitatory element that coordinates glomerular activity. *J Neurosci*, 24(30):6676–6685.

Hildebrand, J. G. (1995). Analysis of chemical signals by nervous systems. *Proc Natl Acad Sci U S A*, 92(1):67–74.

Hildebrand, J. G. and Shepherd, G. M. (1997). Mechanisms of olfactory discrimination: converging evidence for common principles across phyla. *Annu Rev Neurosci*, 20:595–631.

Holt, G. R., Softky, W. R., Koch, C., and Douglas, R. J. (1996). Comparison of discharge variability in vitro and in vivo in cat visual cortex neurons. *J Neurophysiol*, 75(5):1806–1814.

Homberg, U. (2002). Neurotransmitters and neuropeptides in the brain of the locust. *Microsc Res Tech*, 56(3):189–209.

Homberg, U., Montague, R. A., and Hildebrand, J. G. (1988). Anatomy of antenno-cerebral pathways in the brain of the sphinx moth manduca sexta. *Cell Tissue Res*, 254(2):255–281.

Huang, J., Zhang, W., Qiao, W., Hu, A., and Wang, Z. (2010). Functional connectivity and selective odor responses of excitatory local interneurons in drosophila antennal lobe. *Neuron*, 67(6):1021–1033.

Husch, A., Paehler, M., Fusca, D., Paeger, L., and Kloppenburg, P. (2009a). Calcium current diversity in physiologically different local interneuron types of the antennal lobe. *J Neurosci*, 29(3):716–726.

Husch, A., Paehler, M., Fusca, D., Paeger, L., and Kloppenburg, P. (2009b). Distinct electrophysiological properties in subtypes of nonspiking olfactory local interneurons correlate with their cell type-specific ca2+ current profiles. *J Neurophysiol*, 102(5):2834–2845.

Junek, S., Kludt, E., Wolf, F., and Schild, D. (2010). Olfactory coding with patterns of response latencies. *Neuron*, 67:872–884.

Kay, L. M. and Stopfer, M. (2006). Information processing in the olfactory systems of insects and vertebrates. *Semin Cell Dev Biol*, 17(4):433–442.

Kirschner, S., Kleineidam, C. J., Zube, C., Rybak, J., Grünewald, B., and Rössler, W. (2006). Dual olfactory pathway in the honeybee, apis mellifera. *J Comp Neurol*, 499(6):933–952.

Kreissl, S., Strasser, C., and Galizia, C. G. (2010). Allatostatin immunoreactivity in the honeybee brain. *J Comp Neurol*, 518(9):1391–1417.

Krofczik, S., Menzel, R., and Nawrot, M. P. (2009). Rapid odor processing in the honeybee antennal lobe network. *Frontiers in Computational Neuroscience*, 2:1–13.

Kroker, K. (2008). The function of interneurons in the antenal lobe of the honey bee. Master's thesis, Universität Würzburg.

Laurent, G. (1999). A systems perspective on early olfactory coding. *Science*, 286(5440):723–728.

Lei, H., Christensen, T. A., and Hildebrand, J. G. (2002). Local inhibition modulates odor-evoked synchronization of glomerulus-specific output neurons. *Nat Neurosci*, 5(6):557–565.

Lei, H. and Vickers, N. (2008). Central processing of natural odor mixtures in insects. *J Chem Ecol*, 34(7):915–927.

Linster, C. and Cleland, T. A. (2009). Glomerular microcircuits in the olfactory bulb. *Neural Netw*, 22(8):1169–1173.

Linster, C., Sachse, S., and Galizia, C. G. (2005). Computational modeling suggests that response properties rather than spatial position determine connectivity between olfactory glomeruli. *J Neurophysiol*, 93(6):3410–3417.

Lliná, R. R. (1988). The intrinsic electrophysiological properties of mammalian neurons: insights into central nervous system function. *Science*, 242(4886):1654–1664.

MacLeod, K. and Laurent, G. (1996). Distinct mechanisms for synchronization and temporal patterning of odor-encoding neural assemblies. *Science*, 274(5289):976–979.

Malaka, R., Ragg, T., and Hammer, M. (1995). Kinetic models of odor transduction implemented as artificial neural networks. simulations of complex response properties of honeybee olfactory neurons. *Biol Cybern*, 73(3):195–207.

Malun, D. (1991a). Inventory and distribution of synapses of identified uniglomerular projection neurons in the antennal lobe of periplaneta americana. *J Comp Neurol*, 305(2):348–360.

Malun, D. (1991b). Synaptic relationships between GABA-immunoreactive neurons and an identified uniglomerular projection neuron in the antennal lobe of Periplaneta americana: a double-labeling electron microscopic study. *Histochemistry*, 96(3):197–207.

Malun, D., Waldow, U., Kraus, D., and Boeckh, J. (1993). Connections between the deutocerebrum and the protocerebrum, and neuroanatomy of several classes of deutocerebral projection neurons in the brain of male periplaneta americana. *J Comp Neurol*, 329(2):143–162.

Markram, H., Toledo-Rodriguez, M., Wang, Y., Gupta, A., Silberberg, G., and Wu, C. (2004). Interneurons of the neocortical inhibitory system. *Nat Rev Neurosci*, 5(10):793–807.

Matsumoto, S. G. and Hildebrand, J. G. (1981). Olfactory mechanisms in the moth manduca sexta: Response characteristics and morphology of central neurons in the antennal lobes. *Proc R Soc Lond*, 213:249–277.

McCormick, D. A., Connors, B. W., Lighthall, J. W., and Prince, D. A. (1985). Comparative electrophysiology of pyramidal and sparsely spiny stellate neurons of the neocortex. *J Neurophysiol*, 54(4):782–806.

Meier, R., Egert, U., Aertsen, A., and Nawrot, M. P. (2008). Find–a unified framework for neural data analysis. *Neural Netw*, 21(8):1085–1093.

Menzel, R., Hammer, M., Müller, U., and Rosenboom, H. (1996). Behavioral, neural and cellular components underlying olfactory learning in the honeybee. *J Physiol Paris*, 90(5-6):395–398.

Mobbs, P. G. (1982). The brain of the honeybee apis mellifera i. the connections and spatial organization of the mushroom bodies. *Philos Trans R Soc Lond B*, 298:309–354.

Mori, K. and Shepherd, G. M. (1994). Emerging principles of molecular signal processing by mitral/tufted cells in the olfactory bulb. *Semin Cell Biol*, 5(1):65–74.

Müller, D., Abel, R., Brandt, R., Zöckler, M., and Menzel, R. (2002). Differential parallel processing of olfactory information in the honeybee, apis mellifera l. *J Comp Physiol A Neuroethol Sens Neural Behav Physiol*, 188(5):359–370.

Nässel, D. R. (1999). Histamine in the brain of insects: a review. *Microsc Res Tech*, 44(2-3):121–136.

Nässel, D. R. and Homberg, U. (2006). Neuropeptides in interneurons of the insect brain. *Cell Tissue Res*, 326(1):1–24.

Nawrot, M., Aertsen, A., and Rotter, S. (1999). Single-trial estimation of neuronal firing rates: from single-neuron spike trains to population activity. *J Neurosci Methods*, 94(1):81–92.

Nawrot, M. P. (2010). *Analysis of Parallel Spike Trains*, chapter Analysis and Interpretation of Interval and Count Variability in Neural Spike Trains., pages 37 – 58. Springer, New York.

Nawrot, M. P., Aertsen, A., and Rotter, S. (2003). Elimination of response latency variability in neuronal spike trains. *Biol Cybern*, 88(5):321–334.

Olsen, S. R., Bhandawat, V., and Wilson, R. I. (2010). Divisive normalization in olfactory population codes. *Neuron*, 66(2):287–299.

Olsen, S. R. and Wilson, R. I. (2008). Lateral presynaptic inhibition mediates gain control in an olfactory circuit. *Nature*, 452(7190):956–960.

Omata, A., Yomogida, K., and Nakamura, S. (1990). Volatile components of apple flowers. *Flavour and Fragrance Journal*, 5:19–22.

Parra, P., Gulyás, A. I., and Miles, R. (1998). How many subtypes of inhibitory cells in the hippocampus? *Neuron*, 20(5):983–993.

Parzen, E. (1962). On estimation of a probability density function and mode. *Ann of Math Stat*, 33:1065–76.

Pearson, K. (1901). On lines and planes of closest fit to systems of points in space. *Philosophical Magazine*, 2 (6):559–572.

Perez-Orive, J., Mazor, O., Turner, G. C., Cassenaer, S., Wilson, R. I., and Laurent, G. (2002). Oscillations and sparsening of odor representations in the mushroom body. *Science*, 297(5580):359–365.

PING, P. I. N. G., Ascoli, G. A., Alonso-Nanclares, L., Anderson, S. A., Barrionuevo, G., Benavides-Piccione, R., Burkhalter, A., Buzsáki, G., Cauli, B., DeFelipe, J., Fairén, A., Feldmeyer, D., Fishell, G., Fregnac, Y., Freund, T. F., Gardner, D., Gardner, E. P., Goldberg, J. H., Helmstaedter, M., Hestrin, S., Karube, F., Kisvárday, Z. F., Lambolez, B., Lewis, D. A., Marin, O., Markram, H., Muñoz, A., Packer, A., Petersen, C. C. H., Rockland, K. S., Rossier, J., Rudy, B., Somogyi, P., Staiger, J. F., Tamas, G., Thomson, A. M., Toledo-Rodriguez, M., Wang, Y., West, D. C., and Yuste, R. (2008). Petilla terminology: nomenclature of features of gabaergic interneurons of the cerebral cortex. *Nar Rev Neurosci*, 9:557–568.

Ponce-Alvarez, A., Kilavik, B. E., and Rhiele, A. (2010). Comparison of local measures of spike time irregularity and relating variability to firing rate in motor cortical neurons. *J Comput Neurosci*, 29:351–365.

Pouzat, C. and Chaffiol, A. (2009). Automatic spike train analysis and report generation. an implementation with r, r2html and star. *J Neurosci Methods*, 181(1):119–144.

Pouzat, C., Delescluse, M., Viot, P., and Diebolt, J. (2004). Improved spike-sorting by modeling firing statistics and burst-dependent spike amplitude attenuation: a markov chain monte carlo approach. *J Neurophysiol*, 91(6):2910–2928.

Rall, W. and Shepherd, G. M. (1968). Theoretical reconstruction of field potentials and dendrodendritic synaptic interactions in olfactory bulb. *J Neurophysiol*, 31(6):884–915.

Robertson, H. M. and Wanner, K. W. (2006). The chemoreceptor superfamily in the honey bee, apis mellifera: expansion of the odorant, but not gustatory, receptor family. *Genome Res*, 16(11):1395–1403.

Root, C. M., Semmelhack, J. L., Wong, A. M., Flores, J., and Wang, J. W. (2007). Propagation of olfactory information in drosophila. *Proc Natl Acad Sci U S A*, 104(28):11826–11831.

Sachse, S. and Galizia, C. G. (2002). Role of inhibition for temporal and spatial odor representation in olfactory output neurons: a calcium imaging study. *J Neurophysiol*, 87(2):1106–1117.

Sachse, S. and Galizia, C. G. (2003). The coding of odour-intensity in the honeybee antennal lobe: local computation optimizes odour representation. *Eur J Neurosci*, 18(8):2119–2132.

Sachse, S. and Galizia, C. G. (2006). Topography and dynamics of the olfactory system. in microcircuits: The interface between neurons and global brain function. In *Dahlem Workshop Report 93*. MIT Press.

Sachse, S., Peele, P., Silbering, A. F., GÃ¼hmann, M., and Galizia, C. G. (2006). Role of histamine as a putative inhibitory transmitter in the honeybee antennal lobe. *Front Zool*, 3:22.

Sandoz, J.-C., Deisig, N., de Brito Sanchez, M. G., and Giurfa, M. (2007). Understanding the logics of pheromone processing in the honeybee brain: from labeled-lines to across-fiber patterns. *Front Behav Neurosci*, 1:5.

Sato, K. and Touhara, K. (2009). Insect olfaction: receptors, signal transduction, and behavior. *Results Probl Cell Differ*, 47:121–138.

Savitzky, A. and Golay, M. (1964). Smoothing and differentiation of data by simplified least squares procedures. *Anal. Chem.*, 36:1627–1639.

Schäfer, S. and Bicker, G. (1986). Distribution of GABA-like immunoreactivity in the brain of the honeybee. *J Comp Neurol*, 246(3):287–300.

Seki, Y. and Kanzaki, R. (2008). Comprehensive morphological identification and gaba immunocytochemistry of antennal lobe local interneurons in bombyx mori. *J Comp Neurol*, 506(1):93–107.

Seki, Y., Rybak, J., Wicher, D., Sachse, S., and Hansson, B. S. (2010). Physiological and morphological characterization of local interneurons in the drosophila antennal lobe. *J Neurophysiol*.

Shang, Y., Claridge-Chang, A., Sjulson, L., Pypaert, M., and MiesenbÃ¶ck, G. (2007). Excitatory local circuits and their implications for olfactory processing in the fly antennal lobe. *Cell*, 128(3):601–612.

Shepherd, G. M. (1963). Neuronal systems controlling mitral cell excitability. *J Physiol*, 168:101–117.

Shepherd, G. M., Chen, W. R., Willhite, D., Migliore, M., and Greer, C. A. (2007). The olfactory granule cell: from classical enigma to central role in olfactory processing. *Brain Res Rev*, 55(2):373–382.

Shinomoto, S. (2010). *Analysis of Parallel Spike Train Data*, chapter Estimating the firing rate, pages 21–37. Springer, New York.

Silbering, A. F. and Galizia, C. G. (2007). Processing of odor mixtures in the drosophila antennal lobe reveals both global inhibition and glomerulus-specific interactions. *J Neurosci*, 27(44):11966–11977.

Stocker, R. F., Lienhard, M. C., Borst, A., and Fischbach, K. F. (1990). Neuronal architecture of the antennal lobe in drosophila melanogaster. *Cell Tissue Res*, 262(1):9–34.

Stopfer, M., Bhagavan, S., Smith, B. H., and Laurent, G. (1997). Impaired odour discrimination on desynchronization of odour-encoding neural assemblies. *Nature*, 390(6655):70–74.

Stopfer, M., Jayaraman, V., and Laurent, G. (2003). Intensity versus identity coding in an olfactory system. *Neuron*, 39(6):991–1004.

Sun, X.-J., Fonta, C., and Masson, C. (1993). Odour quality processing by bee antennal lobe interneurones. *Chem. Senses*, 18(4):355–377.

Suzuki, H. (1975). Antennal movements induced by odour and central projection of the antennal neurones in the honey-bee. *J Insect Physiol*, 21(4):831 – 847.

Tanaka, N. K., Ito, K., and Stopfer, M. (2009). Odor-evoked neural oscillations in drosophila are mediated by widely branching interneurons. *J Neurosci*, 29(26):8595–8603.

Thorndike, R. L. (1953). Who belongs in the family? *Psychometrika*, 18.

Tollsten, L. and Knudsen, J. T. (1992). scent in dioecious salix (salicaceaae) - a cue determining the pollination system? *Plant Systematics and Evolution*, 182:229–237.

Urban, N. N. and Arevian, A. C. (2009). Computing with dendrodendritic synapses in the olfactory bulb. *Ann N Y Acad Sci*, 1170:264–269.

van Essen, D. (2003). *Visual Neuroscience*, chapter Organization of Visual Areas in Macaque and Human Cerebral Cortex, pages 507 – 521. MIT Press.

Vetter, R. S., Sage, A. E., Justus, K. A., CardÃ©, R. T., and Galizia, C. G. (2006). Temporal integrity of an airborne odor stimulus is greatly affected by physical aspects of the odor delivery system. *Chem Senses*, 31(4):359–369.

Vosshall, L. B., Wong, A. M., and Axel, R. (2000). An olfactory sensory map in the fly brain. *Cell*, 102(2):147–159.

Ward, J. H. (1963). Hierachical grouping to optimize an objective function. *Journal of the American Statistic Association*, 58:236–244.

Willhite, D. C., Nguyen, K. T., Masurkar, A. V., Greer, C. A., Shepherd, G. M., and Chen, W. R. (2006). Viral tracing identifies distributed columnar organization in the olfactory bulb. *Proc Natl Acad Sci U S A*, 103(33):12592–12597.

Wilson, R. I. and Laurent, G. (2005). Role of GABAergic inhibition in shaping odor-evoked spatiotemporal patterns in the Drosophila antennal lobe. *J Neurosci*, 25(40):9069–9079.

Wilson, R. I. and Mainen, Z. F. (2006). Early events in olfactory processing. *Annu Rev Neurosci*, 29:163–201.

Wilson, R. I., Turner, G. C., and Laurent, G. (2004). Transformation of olfactory representations in the drosophila antennal lobe. *Science*, 303(5656):366–370.

Winnington, A. P., Napper, R. M., and Mercer, A. R. (1996). Structural plasticity of identified glomeruli in the antennal lobes of the adult worker honey bee. *J Comp Neurol*, 365(3):479–490.

Yaksi, E. and Wilson, R. I. (2010). Electrical coupling between olfactory glomeruli. *Neuron*, 67(6):1034–1047.

Yuste, R. (2005). Origin and classification of neocortical interneurons. *Neuron*, 48(4):524–527.

i want morebooks!

Buy your books fast and straightforward online - at one of world's fastest growing online book stores! Environmentally sound due to Print-on-Demand technologies.

Buy your books online at
www.get-morebooks.com

Kaufen Sie Ihre Bücher schnell und unkompliziert online – auf einer der am schnellsten wachsenden Buchhandelsplattformen weltweit! Dank Print-On-Demand umwelt- und ressourcenschonend produziert.

Bücher schneller online kaufen
www.morebooks.de

 VDM Verlagsservicegesellschaft mbH
Heinrich-Böcking-Str. 6-8   Telefon: +49 681 3720 174   info@vdm-vsg.de
D - 66121 Saarbrücken   Telefax: +49 681 3720 1749 www.vdm-vsg.de

Printed by Books on Demand GmbH, Norderstedt / Germany